T0127654

Lecture Notes in Artificial Intelligence 9501

Subseries of Lecture Notes in Computer Science

LNAI Series Editors

Randy Goebel
University of Alberta, Edmonton, Canada
Yuzuru Tanaka
Hokkaido University, Sapporo, Japan
Wolfgang Wahlster
DFKI and Saarland University, Saarbrücken, Germany

LNAI Founding Series Editor

Joerg Siekmann
DFKI and Saarland University, Saarbrücken, Germany

More information about this series at http://www.springer.com/series/1244

Madalina Croitoru · Pierre Marquis
Sebastian Rudolph · Gem Stapleton (Eds.)

Graph Structures
for Knowledge Representation
and Reasoning

4th International Workshop, GKR 2015
Buenos Aires, Argentina, July 25, 2015
Revised Selected Papers

 Springer

Editors
Madalina Croitoru
LIRMM
Montpellier Cedex 5
France

Pierre Marquis
CRIL-CNRS
Lens
France

Sebastian Rudolph
Fakultät Informatik
Technische Universität Dresden
Dresden, Sachsen
Germany

Gem Stapleton
University of Brighton
Brighton
UK

ISSN 0302-9743   ISSN 1611-3349 (electronic)
Lecture Notes in Artificial Intelligence
ISBN 978-3-319-28701-0   ISBN 978-3-319-28702-7 (eBook)
DOI 10.1007/978-3-319-28702-7

Library of Congress Control Number: 2015958918

LNCS Sublibrary: SL7 – Artificial Intelligence

Printed on acid-free paper

This Springer imprint is published by SpringerNature
The registered company is Springer International Publishing AG Switzerland

Preface

Versatile and effective techniques for knowledge representation and reasoning (KRR) are essential for the development of successful intelligent systems. Many representatives of next-generation KRR systems are founded on graph-based knowledge representation formalisms and leverage graph-theoretical notions and results. The goal of the workshop series on Graph Structures for Knowledge Representation and Reasoning (GKR) is to bring together researchers involved in the development and application of graph-based knowledge representation formalisms and reasoning techniques.

This volume contains the revised selected papers of the fourth edition of GKR, which took place in Buenos Aires, Argentina, on July 25, 2015. Like the previous editions, held in Pasadena, USA (2009), Barcelona, Spain (2011), and Beijing, China (2013), the workshop was associated with IJCAI (the International Joint Conference on Artificial Intelligence), thus providing the perfect venue for a rich and valuable exchange.

The scientific program of this workshop included many topics related to graph-based knowledge representation and reasoning such as argumentation, conceptual graphs, RDF, representations of constraint satisfaction problems and many more. All in all, the fourth edition of the GKR workshop was very successful. The papers coming from diverse fields all addressed various issues for knowledge representation and reasoning and the common graph-theoretic background allowed to bridge the gap between the different communities. This made it possible for the participants to gain new insights and inspiration.

We are grateful for the support of IJCAI and we would also like to thank the Program Committee of the workshop for their hard work in reviewing papers and providing valuable guidance to the contributors. But, of course, GKR 2015 would not have been possible without the dedicated involvement of the contributing authors and participants.

November 2015

Madalina Croitoru
Pierre Marquis
Sebastian Rudolph
Gem Stapleton

Organization

Workshop Chairs

Madalina Croitoru   LIRMM, Université Montpellier II, France
Pierre Marquis   CRIL-CNRS and Université d'Artois, France
Sebastian Rudolph   Technische Universität Dresden, Germany
Gem Stapleton   University of Brighton, UK

Program Committee

Alessandro Antonucci   IDSIA, Switzerland
Abdallah Arioua   INRA, LIRMM, Université Montpellier II, France
Radim Belohlavek   Palacky University, Olomouc, Czech Republic
Zied Bouraoui   CRIL CNRS UMR 8188, France
Dan Corbett   Optimodal Technologies, USA
Olivier Corby   Inria, France
Cornelius Croitoru   University Al.I.Cuza Iaşi, Romania
Frithjof Dau   SAP, Germany
Juliette Dibie-Barthélemy   AgroParisTech, France
Catherine Faron Zucker   I3S, UNS-CNRS, France
Sebastien Ferre   Université de Rennes 1, France
Christophe Gonzales   LIP6, Université Paris 6, France
Ollivier Haemmerlé   IRIT, Université Toulouse le Mirail, France
Nathalie Hernandez   IRIT, France
John Howse   University of Brighton, UK
Robert Jäschke   L3S Research Center, Germany
Jérôme Lang   LAMSADE, France
Nir Oren   University of Aberdeen, UK
Nathalie Pernelle   LRI-Université Paris Sud, France
Simon Polovina   Sheffield Hallam University, UK
Eric Salvat   IMERIR, France
Fatiha Saïs   LRI and Inria Saclay, France
Nic Wilson   4C, UCC, Cork, Ireland
Stefan Woltran   Vienna University of Technology, Austria

Additional Reviewer

Peter Chapman   University of Brighton, UK

Contents

Designing a Knowledge Representation Tool for Subject Matter Structuring

Giovanni Adorni[✉] and Frosina Koceva[✉]

University of Genoa, Genova, Italy
adorni@unige.it, frosina.koceva@edu.unige.it

Abstract. Relying on pedagogical theories of subject matter structuring and presentation, this paper focuses on the design of a knowledge representation tool for the scheming and organization of educational materials. The idea originates from the Educational Concept Maps model - logical and abstract annotation system, developed with the aim of guaranteeing the reusability of both teaching materials and knowledge structures; in fact, the knowledge structure could be reused for design of different courses according to the learner target. A sequence of concepts characterizing the subject matter under design (lesson or entire course) define a teaching/learning path through the map. It represents the output of the design process of lesson plan, which could be imported in a text-editor, in a LCMS, or presented as web pages. The final goal is to develop a tool assisting the teacher in the daily design of lesson plans via a pliable structured model of domain knowledge.

Keywords: Knowledge representation · Knowledge management · Instructional design · Topic maps

1 Introduction

A student facing a computer terminal reads/listens materials, answers questions or solve problems displayed on the screen by manipulating the keyboard or through multimedia interface. This is a common picture in the case of e-learning processes but which can also be representative of cases of the wider use of digital technologies in everyday teaching practice at school. The questions/problems probe the student's knowledge, skills and competences in a scholarly subject (such as arithmetic, history, grammar, foreign language, etc.), and because of this assessment, the computer proceeds to teach didactic materials to the student.

In this scenario, didactic materials play a key role as well as how such materials are organized and structured.

Between the different theories available in the literature (see, for example, [1, 2]), *Knowledge Space Theory (KST)* [3] allows the representation of a network of possible *Knowledge States* organizing the relevant scholarly subject (i.e., the Subject Matter).

It is worth noting that a Knowledge State is not a quantitative estimate of how much a student has learned. Rather, it is a precise description of what the student knows and does not know at a given moment. According to this theory, a student's competence in

© Springer International Publishing Switzerland 2015
M. Croitoru et al. (Eds.): GKR 2015, LNAI 9501, pp. 1–14, 2015.
DOI: 10.1007/978-3-319-28702-7_1

the subject matter at a given instant of her/his learning process may be identified with one of these states, which is a collection of the concepts (or skills, facts, problem-solving methods, etc.) that the student has mastered.

Learning Space Theory (LST) is a special case of KST where the Knowledge State of a student reveals exactly what she/he is 'ready to learn' (see [4] for a discussion on this topic); such state is then represented by the subset of items in the domain that she/he is capable of answering correctly.

The collection of all the Knowledge States establish a *Knowledge Space* that, by convention, always contains at least (1) the empty state, that is, the state with no items, corresponding to a student knowing nothing in the subject matter, or (2) the entire domain, which is the state of a student knowing everything in the subject matter.

While LST is mainly focused on assessment aspect, which is to gauge the knowledge state of the student as accurately as possible, in this paper we face the problem of organize and structure the subject matter discussing a model, *Educational Concept Map* [5], and an instructional authoring system based on that model. Such system, called *ENCODE – ENvironment for COntent Design and Editing,* integrates two different models into a same environment. The pedagogical model is that presented by Stelzer and Kingsley in "Theory of Subject Matter Structure" [6]. The reference model for the representation of the subject matter is that of subject centric networks with specific focus on Topic Maps [7].

The rest of the paper is structured as follows. Next section is a surfing on the literature of reference; Sect. 3 introduce the Educational Concept Map model; Sect. 4 presents the architecture of the ENCODE system. Section 5 illustrates how the Educational Concept Map model has been implemented by means of Topic Maps. In the last section we report same concluding remarks.

2   Surfing the Literature

Since their origins, Knowledge Representation theories have been applied in the context of *Computer-Assisted Instruction* with the aim of supporting *Intelligent Tutoring Systems* (see for example [8, 9]) by means of a formal representation of: the subject matter to be taught; the educational goals; the starting and final level of knowledge and competencies; and the learning model [10].

Concerning the first issue, Stelzer and Kingsley [6] proposed a comprehensive theory for organizing and formally describing subject matter structures. In their theory, founded on the paradigm of axiomatics, these structures are composed by *content* (content elements consisting of *Primary Notions, Secondary Notions, Basic Principles* and *Established Principles*) and *tasks*. Moreover, they introduced the notion of *dependency* between content and task components to restrict the order in which contents can be presented in the course of the learning processes.

The new millennium has seen a revamp of interest for those studies, related to the representation of knowledge content structures in e-learning context. The major contri-bution to these topics has come from the studies on *Educational Modeling Languages* (EMLs) [11] and on the *Semantic Web* field [12].

EMLs provide a formal conceptualization of the learning process describing *learning units* to promote the reuse and the exchange of these descriptions among different e-learning environments. Martínez-Ortiz and colleagues [13] classified EMLs into the following categories: *evaluation languages*; *content structuring languages*; and, finally, *activity languages*. Among the "content structuring languages" the following can be identified: *Learning Material Markup Language* (LMML) by University of Passau [14]; *TArgeted Reuse and GEneration of TEAching Materials* (Targeteam) by Universität der Bundeswehr [15]; *AICC Course Structure Data Model* [16]; *ADL Sharable Content Object Reference Model 2004* [17].

LMML is based on the *Passau Teachware Metamodel* that offers a general description of the modular structure of educational contents. According to it, educational contents are organized into a hierarchy of *Modules*. The Passau Teachware Metamodel supports predefined associations, such as *defines*, *illustrates*, and *references*. Moreover, prerequisites can be expressed using corresponding associations among modules and, with tasks and objectives, can be incorporated using appropriate meta-data.

The Targeteam language [15], built around the XML-based language *TeachML*, enables the management of knowledge contents in a structured and interrelated way. It is based on the following assumptions: a simple linear sequence of learning contents is not sufficient to support meaningful learning processes; and, consequently, the subject matter structure is crucial for providing an understanding for learners. TeachML consists of several XML sublanguages for different purposes; to support: an abstract structuring of contents, the integration and adaptation of other modules, references to other modules, and different kinds of contents. It is worth noting that LMML and Targeteam-TeachML do not enable instructional designers to create different paths throughout the course [18].

AICC [16] provides a description of sequencing in a course structure. The parts of the course that can be rearranged to define the course order and structure (organizing it into logical sections or units) are referred to as the following structure elements: *assignable units* (AU, also referred to as a lesson) and the *blocks* (a group of lessons and other blocks). Another element to be considered in defining the prerequisites for a course is the *objective* that can be associated with AUs and Blocks. The sequencing within a course is controlled using *prerequisites,* a set of requirements that must be satisfied by a student before entering a new AU.

ADL [17] defines the technical foundations of a Web-based Learning Management System; In 2009 it has released SCORM 2004 4th edition. According to it, the subject matter is organized into a hierarchical structure. Each learner's experience with the same knowledge content structure may be different, depending on the sequencing information that was defined by the content developer and on the learner's specific interactions with content objects.

Felix and Paloma [19] presented a framework based on an instructional application model, called *Xedu*, that provides entities representing instructional components and that will drive the instructional design process.

Pecheanu et al. [20] proposed a theoretical model, named COUL-M (COnceptual Units' Lattice Model), for representing the knowledge domain of a given instructional system. The main elements of their model are the *Conceptual Unit*, a group of related concepts belonging to the knowledge domain; the *Conceptual Structure*, and the

Conceptual Transition Path, a set of conceptual units interconnected within a Conceptual Structure. To apply this model, an instructional designer should start from an analysis of the pedagogical relations among domain's concepts, such as the *Precedence Relation* which concerns the order of concepts within a course; the *Contribution Relation* which concerns the explicit contribution of a concept in the presentation of another concept.

As previously said, a great contribution to the problem of knowledge structure representation in learning design has come from the field of Semantic Web (typically, such researches are based on an ontological approach to the conceptualization process – see for example [21, 22]).

Devedžić [23] classifies educational ontology into *domain ontology* (describing essential concepts, relations, and theories of a specific domain, not necessary only for educational purposes), *task ontology* (specifying features of problem solving processes), *teaching strategy ontology* (describing the knowledge and principles underlying pedagogical actions and behaviors), *learner model ontology* (representing learner's performances and cognitive traits), *interface ontology* (specifying e-learning system's adaptive behaviors and techniques at the user interface level), *communication ontology* (defining the semantics of message content languages), and, finally, *educational service ontology* (providing a semantic description of educational services and the underlying logic).

It is interesting to notice that none of these categories concerns the representation of learning content structure (and learning goals), regardless of its knowledge domain. A *learning content structure ontology* should be based on several kinds of links that represent different types of relationships among subject matter topics; the most common associations in these ontologies are *"prerequisite links"*, representing the fact that a given concept has to be learned before another one. Other common relationships are traditional semantic links, such as *"is-a"* and *"part-of"* (categorizing topics into classes and subclasses). In more elaborate ontologies, the formal representation can include a vocabulary to classify educational units, such as *definition*, *example*, or *exercise* [22].

Stojanovic et al. [24], referring to ontology-based metadata in the context of e-learning environments, discuss the use of metadata for describing the structure of learning materials. In this respect, several types of structuring relations between chunks of learning material may be identified, such as *Prev, Next, IsPartOf, HasPart, References, IsReferencedBy, IsBasedOn, IsBasisFor, Requires, IsRequiredBy*. On the other hand, it is interesting to put in evidence that such a kind of metadata representing relationships also appears in conventional metadata models, such as the IEEE Learning Object Metadata at the *Relation* level [25].

An example of content structure ontology, related to the learning domain, has been proposed by Jovanović et al. [26] whose approach is based on the *Abstract Learning Object Content Model* (ALOCoM) ontology. This latter defines concepts and relationships that enable formal definition of the structure of a learning object.

3 Educational Concept Maps

3.1 Overview

The goal of this work is to develop a system that assists the teacher for the design of a course structure, i.e. a network of Knowledge States (nodes), and related teaching materials associated to each state (node), with the aim of guaranteeing the reusability of both network and materials. The network of knowledge states can be of two types: an *ECM – Educational Concept Map*, representing a general structure of the domain of knowledge, or a *CCM - Course Concept Map*, representing an instance of that domain.

An ECM is a formal representation of the domain of knowledge in the context of learning environments. It is a logical and abstract annotation model created with the aim of guaranteeing the reusability of the knowledge structure, as well as of the teaching materials associated to each node of the structure. ECM model has been designed taking into account the pedagogical requirements defined by Educational Modeling Language research group [27], namely:

1. pedagogical flexibility: the model must be able to describe the structure of instructional contents regardless of a specific learning theory;
2. learner centrality: the instructional content design process must be based on students' profile and needs;
3. centrality of learning objectives: the instructional content design process must be based on a preliminary definition of learners' pedagogical objectives;
4. personalization: the model must be able to design learning contents and resources in a flexible way, consistently with learners' profile;
5. domain-independence: the model must be able to describe instructional content regardless of its disciplinary nature;
6. reusability: the model must allow to define and de-contextualize learning content structures and to reuse these in other contexts;
7. interoperability: the model must be language-independent, so that it can be implemented through different knowledge representation languages and exported in different learning environments;
8. medium neutrality: the instructional contents design process must be medium neutral, so that it can be used in different publication formats;
9. compatibility: the model must fit in with existing standards and specifications on learning resources;
10. formalization: the model must describe instructional content according to a formalized model, so that automated processing is possible.

Figure 1 represents a high level view of the activity diagram of the ENCODE system.

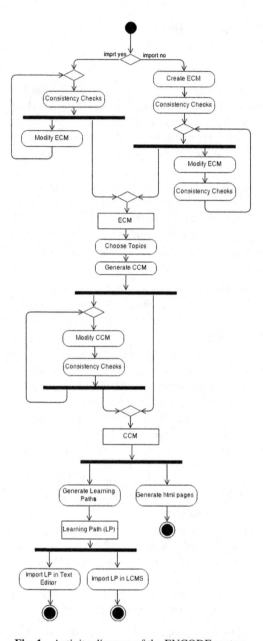

Fig. 1. Activity diagram of the ENCODE system

Where a user may work on a previously defined ECM (i.e., import) or may "create" a new structure of a domain knowledge (i.e., a new ECM), and then, an instance of that domain (i.e., a CCM). Throughout the following sections we will describe the activity of that figure.

3.2 ECM Model Structure

The ECM model has been developed by means of an ontological structure characterized by the integration of hierarchical and associative relationships. Firstly, it asks teachers and instructional designers to focus their attention on learners' profile (in particular educational background, learning and cognitive styles) and objectives. Taking into account these elements, the model suggests how to identify, within the discipline's subject matter, the key concepts and their relationships to identify effective strategies of contents presentation and to support the activation of meaningful learning processes.

According to the ECM model (see Fig. 2), a profiled learner has a Lesson Plan with a goal identified by an objective (or a composition of objectives) that is (are) achieved by a *Unit of Learning (UoL)*, or by a composition of UoLs. A UoL is characterize by an Effort, i.e., an evaluation of the needed commitment that the learner requires in dealing with the learning materials. A UoL is composed by key concept and their relationships, where the key concepts can be of two types:

- *Primary Notion Type - PNT* when the concept is the starting point of a UoL and identifies the "prerequisites", i.e., the concepts that a student must know before attending a given UoL.
- *Secondary Notion Type - SNT* identifies the concepts that will be explained in the present UoL (this kind of concepts go with learning materials).

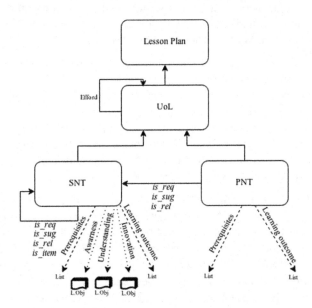

Fig. 2. The ECM model

The ECM model has been developed by means of an ontological structure characterized by the integration of hierarchical and associative relationships. Thus, four educational relations are defined:

- *is-requirement-of*: denoted as *is_req (x, y)* a propaedeutic relation, e.g., it may be used with the aim of specifying the logical order of contents, i.e., used for creating learning linearized paths;
- *is-item-of*: denoted as *is_item (x, y)*, representing hierarchical structure of an aggregation or membership type;
- *is-related-to*: denoted as *is_rel (x, y)*, represents a relation between closely related concepts;
- *is-suggested-link-of*: denoted as *is_sug (x, y)*, relates a main concept with its in-depth examination, e.g., this relationship type may be used in order to suggest in-depth resources.

A CCM is the structure of the subject matter, e.g., a specific teacher vision of the course domain tailored for a specific class of students. As to reusability, the ECMs are designed to maintain the concept layer separate from the resources, making it possible to provide courses with the same CCM from the ECM but with different resources, as in the case of a course for beginners and a course on the same argument for advanced students.

4   The ENCODE Architecture

The ECM model is the theoretical framework for the design of ENCODE application, currently in the implementation phase, with some innovative features described in the following of this section. Starting from an ECM a user can:

- publish a CCM on the Web, where the relationships suggest the different navigation strategies of the underlying subject matter.
- generate a linearized path, useful, for example, for a teacher to produce a lesson or a document about a given subject matter. In this latter case, a *Suggested Paths Strategy* is necessary, to be expressed by means of *is_req* relationships. The linearized path will be than used to create didactic materials on such subject (by means of a text editor) or to generate a course through a Learning Content Management System (LCMS) as explained at the end of this section.

To explain the approach behind the Suggested Paths Strategy, let us also consider the idea of preparing a lesson on a given subject matter, using the previous ECM model. The *is_req* relationships order the topics T of the lesson according to the propaedeutics rules, therefore in the graph $G = (T, E)$ there cannot be loops, thus obtaining a Direct Acyclic Graph, where T are nodes and E arcs, with: $(t_i, t_j) \in E \leftrightarrow is\_req(ti, tj)$. In this context, a *Topological Order* on a CCM is a sequence $S = \{s_1, s_2, \ldots s_{|T|}\}$ where each element T appears only once and cannot be preceded by any of its successors; given pair of nodes (t_i, t_j) in S if there exists an arc from t_i to t_j of type is_req, it follows that the node t_i is before the node t_j in the list: $\forall(t_i, t_j) \in S: (t_i, t_j) \in E \rightarrow i < j$.

The algorithm implementing the Topological Order is derived by Topological sorting algorithm [28] with a main modification in order to get all the possible sequences of topological sorting. The possible sequences are obtained by permutations between topics without invalidating the prerequisite constraint. Therefore we let the teacher to choose which of these sequences better answers the accomplishment of the didactic objectives. The result of topological ordering is a XML structure.

For as much the topics are topologically ordered this doesn't take into account the distance factor in between the topics, thus a signaling (denoted as *Topic Aider - TA*) is introduced in the sequence S before the distant topic to remind its subject. The TA is a suggestion for the teacher to introduce an exercise, an example, a text or a valuation test. This TA is also reported in the final sequence in order to highlight not only to the teacher, but also to the student the place where s/he should evoke a determinate argument. The choice to have not a single path but a list of paths to suggest to the author leaving the final choice to the author him/herself, is also to answer to the non-equifinality problem posed in [29]. The "suggested" order lists is on the basis of the principle of reducing as much as possible the distance between two topics of the list that are contiguous on the graph.

Furthermore, for better presentation of the knowledge structure and effective navigation a cluster with a name Nc is defined, by grouping all the topics t_i which are in an *is_item* relationship, i.e., *is_item(t_i, t_j)* with a common topic t_j. More formally we define a cluster $C = \{Nc, Tc\}$ as a non-empty finite set of topics Tc, where $\forall t_i \in Tc$ and $\exists t_j \notin Tc$ where *is_item(t_i, t_j)*.

The ENCODE (see Fig. 3) is web based application designed on top of the Ontopia Engine [30] by implementing the specific constraints of the ECM model. Taking into account the target end-user and retaining that graph-based knowledge representation is more intuitively understandable, ENCODE will implement a graph-based interface.

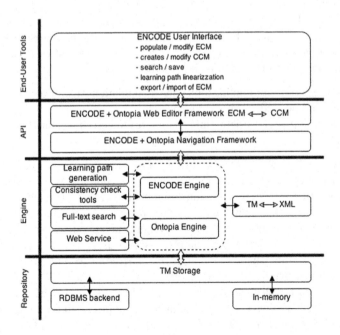

Fig. 3. ENCODE Architecture

Providing a graphical visualization/navigation and editing interface lets the map-designer to incrementally populate the ECM/CCM.

The End-user layer provides also an export possibility of the ECM or of the topological order, i.e., the linearized lesson plan. Thus, for editorial purposes the topological order can be exported in a XML format, which maintains the structure of the lesson plan and makes it possible to be imported in a text editor for further adaptations. The XML format can be imported in LCMS, giving an initial shape of the course and the learning materials. Last but not least, form the XTM format (serialization of the ECM map, see ECM representation) html pages can be generated transforming the propaedeutic order in navigational.

The API Layer is a JSP tag library and Java API for developing a web-based interface based on topic maps. For the storing and providing access to the educational topic maps the underlying layer is used, i.e., the engine layer.

The engine is a set of Java APIs which with the ENCODE extension manages both the ECM; the CCM; provides utilities for the consistency checks, for the generation of learning paths (implementation of the topological ordering algorithm), etc. It can keep maps in memory or store them in a relational database.

5 The ECM Representation

In order to implement the ECM-model, TM has been chosen. TM is an ISO multi-part standard [7] designed for encoding knowledge and connecting this encoded knowledge to relevant information resources. The standard defines a data model for representing knowledge structures and a specific XML-based interchange syntax, called *XML Topic Maps (XTM)* [31]. The main elements in the TM paradigm are: *topic* (a symbol used to represent one, and only one, subject), *association* (a relationship between two or more topics) and *occurrence* (a relationship between a subject and an information resource). Therefore, two layers can be identified into the TMs paradigm:

- The *knowledge layer* representing topics and their relationships, allowing to construct our network of knowledge states (topics);
- The *information layer* describing information resources, to be attached to the ECM topics.

Each topic can be featured by any number of *names* (and *variants* for each name); by any number of *occurrences*, and by its *association role*, that is a representation of the involvement of a subject in a relationship represented by an association. All these features are statements and they have a *scope* representing the context a statement is valid in. Using scopes it is possible to avoid ambiguity about topics; to provide different points of view on the same topic (for example, based on users' profile) and/or to modify each statement depending on users' language, etc. Therefore, to solve ambiguity issues, each subject, represented by a topic, is identified by a *subject identifier*. This unambiguous identification of subjects is also used in TMs to merge topics that, through these identifiers, are known to have the same subject (two topics with the same subject are replaced by a new topic that has the union of the characteristics of the two originals).

The TM data model uses an infoset formalism plus UML diagrams (for illustration purposes) to describe the Topic Maps. We define the design rules of the ECM model via ECM Schema and *tolog* rules, i.e., tolog is a language for querying and updating topic maps. Thus, by defining the constraints we give an educational interpretation of the topic map enhancing it with pedagogical concepts, e.g., prerequisite structure, Primary Notion, Learning Outcomes, Effort. The ECM schema defines a collection of typing topics for our ECM and the specific constraints on them where a constraint defines either a way in which types can be combined in the TM or a way in which a schema type and property values can be combined. One of the limitation in our model in respect of the TM model is that the Association in ECM are binary - involve only two topics, instead of the n-ary association in the TM model. If we denote by T the set of educational topics, and by R the set of types of educational association, for $x, y \in T$ and $x \neq y$, $r(x, y)$ is a generic educational association type member of $R = \{is\_req(x, y), is\_item(x, y), is\_rel(x, y), is\_sug(x, y)\}$. The set R of four education association types constraints the interaction between the topics representing the domain of interest in order to support the designer of the educational map. They have been selected with the aim of allowing teachers to create different learning paths (with or without precedence constraints among topics). More precisely, the association types are:

- *is_req (x, y)* stands for "topic x is a requirement of topic y". Identifies a asymmetric (1), transitive (2) and propaedeutic association between topic x with role prerequisite and y with role subsidiary;
- *is_item (x, y)* identifies a hierarchical asymmetric (4) association among two topics x and y, where x is an individual topic (role individual) and y is its more general topic (role general). It can be used also to identify a membership relationship;
- *is_rel (x, y)* identifies a symmetric (5) association among closely related topics; where topic x is of role linked and topic y of role linked (e.g., it may be used with the aim of creating learning paths without precedence constraints);
- *is_sug (y, x)*, identifies an asymmetric (3) association among a topic y with role main and it's in-depth examination x with role deepening. The deepening topics are leafs in the map (7)

$$\forall x, y \; is\_req(x, y) \rightarrow \neg \; is\_req(y, x) \tag{1}$$

$$\forall x, y \; is\_req(x, y) \wedge is\_req(y, z) \rightarrow is\_req(x, z) \tag{2}$$

$$\forall x, y \; is\_sug(y, x) \rightarrow \neg \; is\_sug(y, x) \tag{3}$$

$$\forall x, y \; is\_item(x, y) \rightarrow \neg \; is\_item(y, x) \tag{4}$$

$$\forall x, y \; is\_rel(x, y) \rightarrow is\_rel(y, x) \tag{5}$$

$$x, y \; is\_sug(y, x) \rightarrow \neg \; \exists z \; [is\_req(x, z) \wedge is\_sug(x, z) \wedge is\_item(x, z) \wedge is\_rel(x, z)] \tag{6}$$

Another constraint of ECM is that two topics x and y can be related only by one type of association (7)

$$\forall x, y \in T \wedge \exists ! \, r(x,y) \in R \qquad\qquad (7)$$

We defined an Education Topic Type, which is a super type of the *PNT* and the *SNT*. The PNT as mentioned is the starting point of a unit of learning, i.e., identifies the concepts that a student must know before attending a given unit of learning. Thus the PNT's will not have learning materials and they do not introduce new knowledge. The SNT instead identifies the concepts that will be explained in the present unit of learning (this kind of topics will have specific learning materials associated by using suitable occurrence type).

By the occurrence types, we represent a special relationship between the topic and its information resource. We define five Occurrence Types to specify the nature of the relationship that the topic has with the information resource: Prerequisites, Learning Outcomes, Awareness, Understanding, and Innovation. The values of the Prerequisite and Learning Outcome occurrence type represents all the prerequisites and learning outcomes of the educational topic. The URLs of the learning materials associated to an education topic are the values of the Awareness, Understanding and Innovation occurrence types. In this manner, we classify the learning materials (usually Learning Objects) in respect of the target of students for the specific CCM. To the learning materials, we also associate a number representing the Effort, e.g., the time effort necessary for learning the material. Once the CCM is created the single efforts are accumulated in the UoL Effort.

The constraint may or may not require the existence of some properties. Thus, all instances of a concept of PNT may take a part of relationship of type is_req with a role prerequisite, or is_sug with a role main, or is_rel with role linked and must have Prerequisite and Learning Outcome occurrence. While all instances of a concept of SNT may take any role in a relationship ($r(x, y) \in R$) and must have Prerequisite and Learning Outcome occurrence and at list one occurrence for the learning materials of type Awareness, Understanding and/or Innovation

In addition to the validation of the map against the ECM schema, we also express design rules for our model by defining specific *tolog* rules. Some examples of constraints of this type are:

- On deletion of a topic *y* where $\exists x, y, z \, is\_req(x, y) \wedge is\_req(y, z)$ first a new association is_req(x, z) is created than a check of existence of a *path(x, z)* is made. If $\exists \, path(x, z)$ than is_req(x, z) is deleted, else is_req(x, z) remains.
- On adding a is_req(x, y) a verification of existence of a *path(x, y)* is made, if $\exists \, path(x, y)$ than the is_req(x, y) is deleted.
- On adding a topic *x* and relating the topic with is_req(x, y) or is_item(x, y), the values of the Learning Outcomes of *x* are "propagated"/added to the list of existing values of Learning Outcomes of *y*, and "propagated"/added to $\forall z \in path(y, z)$.
- On some types of inconsistency, e.g., patterns not found in the ECM, an inconsistency property flag is added.

6   Conclusions

The idea behind this work has been stimulated by the real needs of a community of teachers to have model and tools that facilitates some phases of instructional design. Since the concept representation is independent of its implementation, ECM lends itself for reusability of both teaching materials and knowledge structure. Thus the knowledge structure (network of knowledge states) could be reused for the design of a different course according to the learner target. Moreover, the underlying model, ECM, is grounded on pedagogical reflections. For these reasons we believe that this model will have a good acceptance by the community of teachers we plan to select for the testing phase.

References

1. Puff, C.R.: Memory Organization and Structure. Academic Press, New York (1979)
2. Neath, I., Surprenant, A.M.: Human memory: an Introduction to Research, Data, and Theory, 2nd edn. Wadsworth, Belmont (2003)
3. Doignon, J.P., Falmagne, J.P.: Knowledge Spaces. Springer, Berlin (1999)
4. Falmagne, J.C., Doignon, J.P.: Learning Spaces. Springer, Berlin (2011)
5. Adorni, G., Brondo, D., Vivanet, G.: A formal instructional model based on Concept Maps. J. E-Learn. Knowl. Soc. **5**(3), 33–43 (2009)
6. Stelzer, J., Kingsley, E.: Axiomatics as a paradigm for structuring subject matter. Instr. Sci. **3**(4), 383–450 (1975)
7. ISO/IEC Home of SC34/WG3 Informational Association. http://www.isotopicmaps.org
8. Murray, T.: Authoring intelligent tutoring systems: an analysis of the state of the art. Int. J. of Artif. Intell. Educ. (IJAIED) **10**, 98–129 (1999)
9. Stankov, S., Glavinićc, V., Rosićc, M.: On knowledge representation in an intelligent tutoring system. In: Proceedings of the Fourth International Conference on Intelligent Engineering Systems (INES 2000), Portorož (Slovenia) (2000)
10. Pask, G.: Computer assisted learning and teaching. In: Annett, J., Duke, J. (eds.) Proceedings of Leeds Seminar on Computer Based Learning. NCET, pp. 50–63 (1970)
11. Rawlings, A., van Rosmalen, P., Koper, R., Rodriguez-Artacho, M., Lefrere, P.: Survey of educational modelling languages (EMLs). In: CEN/ISSS WS Learning Technologies Workshop (2002)
12. Berners-Lee, T., Hendler, J., Lassila, O.: The semantic web. Sci. Am. **284**(5), 28–37 (2001)
13. Martínez-Ortiz, I., Moreno-Ger, P., Sierra, J.L., Fernandez-Manjon, B.: Supporting the authoring and operationalization of educational modelling languages. J. Univ. Comput. Sci. **13**(7), 938–947 (2007)
14. Süß, C., Freitag, B.: LMML – the learning material markup language framework. In: Proceedings of International Conference on Interactive Computer Aided learning ICL, Villach (2002)
15. Teege, G.: Documentation of Targeteam V 0.5.8 (2004). http://inf3-www.informatik.unibw-muenchen.de/research/projects/targeteam/targeteam-0.5.8/doc/targeteam.pdf
16. AICC-Aviation Industry CBT Committee - CMI Guidelines for Interoperability. Revision 3.5, Release 2 (2001). http://www.aicc.org/docs/tech/cmi001v3-5.pdf

17. ADL Advanced Distributed Learning - SCORM 2004 4th Edition Version 1.1 Documentation (2009). http://www.adlnet.gov/wp-content/uploads/2011/07/SCORM.2004.3ED.DocSuite.zip

18. Wehner, F., Lorz, A.: Developing modular and adaptable courseware using TeachML. In: Proceedings of 13th ED-MEDIA, World Conference on Educational Multimedia, Hypermedia and Telecommunications, Tampere (2001)

19. Félix, B., Paloma, D.: A framework for the management of digital educational contents conjugating instructional and technical issues. Educ. Technol. Soc. 6(4), 48–59 (2003)

20. Pecheanu, E., Stefanescu, D., Istrate, A., Dascalescu, D.: Conceptually modeling the domain knowledge for assisted learning in an IT discipline. In: Proceedings of International Conference on Computer Aided Learning in Engineering Education, Grenoble, pp. 215–220 (2004)

21. Mizoguchi, R., Bourdeau, J.: Using ontological engineering to overcome common AI-ED problems. Int. J. Artif. Intell. Educ. 11, 107–121 (2000)

22. Pahl, C., Holohan, E.: Applications of semantic web technology to support learning content development. Interdisc. J. E-Learn. Learn. Objects 5, 1–25 (2009). Dublin City University, Ireland

23. Devedžić, V.: Semantic Web and Education. Integrated Series in Information Systems, vol. 12. Springer, US (2006)

24. Stojanovic, L., Staab, S., Studer, R.: E-learning based on the Semantic Web. In: Proceedings of WebNet2001-World Conference on the WWW and Internet, Orlando (2001)

25. IEEE Learning Technology Standards Committee. IEEE Standard for Learning Object Metadata (2002). http://ltsc.ieee.org/wg12/

26. Jovanović, J., Gašević, D., Devedžić, V.: Dynamic assembly of personalized learning content on the semantic web. In: Sure, Y., Domingue, J. (eds.) ESWC 2006. LNCS, vol. 4011, pp. 545–559. Springer, Heidelberg (2006)

27. Koper, R.: Modelling Units of Study from a Pedagogical Perspective: the Pedagogical Metamodel Behind EML. Open Universiteit Nederland, Heerlen (2001)

28. Kahn, A.B.: Topological sorting of large networks. Commun. ACM 5(11), 558–562 (1962)

29. Ohlsson S.: Some principles of intelligent tutoring. In: Lawler, R.W., Masoud Yazdani, M. (eds.) Artificial Intelligence and Education: Learning Environments and Tutorial Systems, vol. 1. Intellect Ltd, Chicago (1987)

30. Ontopia Project. http://www.ontopia.net/

31. Garshol, L.M., Graham, M.: Topic Maps - XML Syntax. Final Draft International Standard. http://www.isotopicmaps.org/sam/sam-xtm/

Aligning Experientially Grounded Ontologies Using Language Games

Michael Anslow[⊠] and Michael Rovatsos

University of Edinburgh, Edinburgh, UK
`M.Anslow@sms.ed.ac.uk, mrovatso@inf.ed.ac.uk`

Abstract. Ontology alignment is essential to enable communication in a multi-agent system where agents have heterogeneous ontologies. We use language games as a decentralised iterative ontology alignment solution in a multi-agent system where ontologies are grounded in measurements taken in a dynamic environment. Rather than attempting to ground ontologies through physical interaction, we design language game strategies that involve exchanging descriptions of the environment as graph patterns and interpreting descriptions using graph matching. These methods rely on structural similarity as evidence for ontology alignment. We compare various language game strategies with respect to communication overhead and alignment success and provide preliminary results which show that ontology alignment using language games that rely on descriptions alone can result in perfect alignments with only modest communication overhead. However, this requires that environmental dynamics are reasoned about when providing descriptions.

1 Introduction

Successful communication in a multi-agent system is paramount to successful coordination. To this end, ontologies make the semantics of a language explicit, in a machine readable form, that facilitate in the interpretation of communication. However, when ontologies are heterogeneous, an alignment between them must be found. In this paper, we explore graph-based ontology matching solutions to this problem, where structural similarity between ontologies serves as evidence for ontology alignment. We do this in the context of a multi-agent simulation where ontologies are grounded in measurements of a dynamic environment.

We adapt 'language games' popularised by [12] as a coordinated communication process that serves as a decentralised, iterative ontology alignment solution. Agents perform communication tasks to distinguish between different elements of their environment by selecting, communicating and reasoning about graph patterns and performing graph matching. Our language games are novel in that they are based on finding overlapping descriptions of a shared environment without relying on alignment through physical interaction alone.

Language games for the purpose of ontological alignment are reduced to three subproblems, *target selection*, *context selection* and *correspondence induction*. Target selection consists of selecting a label that agents will attempt to

© Springer International Publishing Switzerland 2015
M. Croitoru et al. (Eds.): GKR 2015, LNAI 9501, pp. 15–31, 2015.
DOI: 10.1007/978-3-319-28702-7_2

discover an alignment for. Context selection consists of selecting a graph pattern whose structure distinguishes the selected target label from other labels. Correspondence induction consists of inducing correspondences between ontologies based on reasoning about structural similarity between local and communicated knowledge. We use a fixed target selection strategy in this paper and instead focus on context selection and correspondence induction. In particular we explore the quality of different solutions to these problems with respect to correctness of alignments and communication overhead. We believe that graph-based structural similarity can resolve practical, task oriented, ontology alignment problems if we assume a sufficiently distinctive environment, that agents structure their knowledge in the same way, and that agents co-exist within and have overlapping knowledge of, their environment.

We only address instance matching and provide fixed and correct alignments between other elements of ontologies such as concepts and relations. We explore two forms of ontological heterogeneity, coverage and terminological mismatch, as described by [7]. Difference in coverage of ontologies results from perceiving the environment at the same level of granularity, but, as a result of incomplete information, ontologies represent a partial perspective of a dynamic environment. Terminological mismatch is caused by autonomous labelling of instances and is inevitable given that agents experience their environment in isolation and discover entities independently.

The approach detailed in this paper is applicable to any ontology matching problem where structural similarity of ontologies is indicative of similarity between ontological elements. We have not addressed a real-world ontology alignment problem. Instead, we favour well-defined, randomly generated ontology matching problems that require robust solutions with respect to variations among generated ontologies. We regard our approach as a first step towards further principled and methodical exploration of language games as a solution to distributed ontology alignment along the dimensions of ontology heterogeneity, agent behaviour and communication requirements.

We provide preliminary experimental results in a simulated grid-world-like scenario which show that ontology alignment using language games that rely on descriptions alone can result in near perfect alignments. However, this requires that environmental dynamics are reasoned about when providing descriptions and that partial matching of descriptions is used when there are inconsistencies between received descriptions and local knowledge.

The remainder of this paper is structured as follows: In Sect. 2, we contextualise our work against related literature. In Sect. 3, we provide a formal definition of the ontology alignment problem within a multi-agent systems context. In Sect. 4, we describe our proposed solution. Section 5 outlines our experimental methodology. We then compare the performance of various language game strategies in Sect. 6. Finally, we summarise our findings and suggest possible extensions to this work in Sects. 7 and 8, respectively.

2   Related Work

Ontology matching [7] is the automatic/semi-automatic discovery of semantic correspondences between heterogeneous ontologies. Existing agent-based ontology matching approaches involve some form of interaction between agents, where agents negotiate the meaning of the correspondences between ontologies [5]. The ontologies that these techniques are applied to are typically Semantic Web ontologies, as initially described by [8]. In general, both agent-based and non-agent-based approaches to ontology matching rely on the assumption that ontologies are constructed by humans. This allows for a rich plethora of natural language processing tools and resources to be used. However, when ontologies are learnt autonomously from non-human perceptions and labelled in an artificial language, this assumption does not hold.

The 'symbol grounding problem' as described by [10] is the problem of how symbols 'acquire' meaning. In robotics and intelligent systems, this problem is that of grounding symbols in data measured by physical sensors. There are two variants of this: 'physical symbol grounding', described by [14], which consists of individual agents grounding symbols by interacting with the environment and 'social symbol grounding', the multi-agent extension to this described by [2], in which agents negotiate the meaning of independently physically grounded symbols. In this work we explore social symbol grounding at a high level of abstraction without taking account of the complexities of low-level signal processing. We believe that anchoring frameworks, as described by [3], provide a plausible solution to the low-level counterpart of this abstraction and therefore, we assume a mapping from low-level signals to symbolic labels is readily available.

Though our work is influenced by language games, it is distinct from existing work that use this technique in a number of ways:

- Agents do not plan actions to discover overlapping measurements. Instead, agent behaviour is fixed and overlapping measurements are coincidental, instead agents align ontologies by discovering structural similarity between knowledge.
- Language games are driven by a need to communicate. As such, language learning is not an end in itself.
- A shared language is not the result of shared invention of labels. Instead, agents have an existing language before attempting to align ontologies that corresponds to their conceptualisation of the environment.
- Language games typically only focus on strategies to discover equivalences between labels, we also infer disjunction between labels when particular equivalences can be ruled out. Disjunctions are used to constrain interpretation of language in subsequent language games.

Unlike the work of [11] we do not explore feedback from a task for the diagnosis and correction of incorrect alignments. Our techniques are evidence based, and as such, prone to error; solutions to correcting incorrect alignments are beyond the scope of this paper.

3   Formal Framework

Agents $A = \{1, 2, \cdots, n\}$ exist within an environment. Each agent maintains a conceptualisation of their environment as an ontology. We define ontology as follows:

Definition 3.1. *An ontology is a tuple $O = \langle \mathcal{C}, \mathcal{I}, \mathcal{T}, \mathcal{D}, \mathcal{R}, \mathfrak{R} \rangle$ where: $\mathcal{C}$ is the set of concepts (classes), $\mathcal{I}$ is the set of individuals/instances, $\mathcal{T}$ is the set of data types, $\mathcal{D}$ is the set of data values, $\mathcal{R}$ is the set of relations, $\mathfrak{R} : (\mathcal{C} \cup \mathcal{I} \cup \mathcal{T} \cup \mathcal{D})^2 \to \wp(\mathcal{R})$ is a function indicating which binary relations hold among $\mathcal{C}, \mathcal{I}, \mathcal{T}, \mathcal{D}$.*

This is similar to the definition used by [7], however we do not differentiate between particular classes of relations such as subsumption, equivalence and exclusion as our focus is only on the structure of an ontology rather than reasoning about what particular relationships entail.

Example 3.1 describes an ontology and Fig. 1 provides a graphical depiction of this ontology. These are representative fragments of the ontologies used in our experiments described in Sect. 5.

Example 3.1. $\mathcal{C} = \{Location, Agent, C_1\}$, $\mathcal{I} = \{A, B, C, D, E\}$, $\mathcal{R} = \{Connected, InLocation, MemberOf, HasValue, HasType\}$, $\mathcal{T} = \{Boolean\}$. *As $\mathcal{T}$ only contains Boolean, $\mathcal{D} = \{True, False\}$. $\mathfrak{R}$ defines: what concept an instance is a members of ($\mathfrak{R}(D, Agent) = \{MemberOf\}$), which location an instance is in ($\mathfrak{R}(D, C) = \{Inlocation\}$), which locations are connected ($\mathfrak{R}(A, B) = \{Connected\}$), what property an instance has ($\mathfrak{R}(A, True) = \{HasValue\}$) and the type of a data value ($\mathfrak{R}(True, Boolean) = \{HasType\}$).*

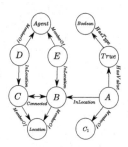

Fig. 1. A graphical depiction of an ontology. This ontology depicts three instances in different locations, two of these instances are members of concept *Agent* and one of an arbitrarily generated concept C_1. The instance A of concept C_1 has a property *True* of type *Boolean*.

The set of possible ontologies is $\mathcal{O}$. The environment, $\mathcal{E}$, is itself an ontology in this set. It is at the abstraction of this ontology that agents perceive, reason and communicate about their environment. Agents 'measure' their environment at each time step according to a measurement function,

$$\mu_i : \mathcal{O} \to \mathcal{O} \tag{1}$$

where $\mu_i(\mathcal{E}) = \hat{\mathcal{E}}$ is the measurement received by agent i. $\hat{\mathcal{E}}$ is a sub ontology of $\mathcal{E}$ such that $\hat{\mathcal{C}} \subseteq \mathcal{C}$, $\hat{\mathcal{I}} \subseteq \mathcal{I}$, $\hat{\mathcal{T}} \subseteq \mathcal{T}$, $\hat{\mathcal{D}} \subseteq \mathcal{D}$ and $\hat{\mathfrak{R}}$ is a sub function of $\mathfrak{R}$. μ defines how complete each agent's measurements are of $\mathcal{E}$, for example, if $\mu_i(\mathcal{E}) = \mathcal{E}$ then agents have complete measurements.

Given measurements, agents independently learn their own conceptualisation of their environment as an ontology, where O_i is the ontology of agent i. To communicate about ontologies, ontologies are associated with a 'labelling' function,

$$\ell : \mathcal{C} \cup \mathcal{I} \cup \mathcal{T} \cup \mathcal{D} \cup \mathcal{R} \to \mathcal{L}, \tag{2}$$

that associates elements of the ontology with a countably infinite set of labels, $\mathcal{L}$.

We assume that the labelling functions for local ontologies and the environment are surjective in general and bijective for instances. Though elements of $\mathcal{D}$ are not uniquely labelled (e.g. *True* and *False* are used for multiple entities), we require that the ontological entities they label can be distinguished from other entities by the entities' relational context. For example, the entity labelled *True* in Fig. 1 can be uniquely identified by its relation with the uniquely labelled instance A.

With these provisions, we can now proceed with our definition of the ontology alignment problem. To facilitate communication between agent i and agent j, agent i must discover an alignment, $\phi \subset \mathcal{L}_i \times \mathcal{L}_j \times \Theta$ where $\mathcal{L}_i$ consists of labels local to agent i, $\mathcal{L}_j$ consists of labels received from j and Θ is a set of semantic relations that can hold between labels. An element of an alignment is called a *correspondence* and is a tuple $\langle l, l', \theta \rangle$ that implies that θ holds between l and l'. We use Φ to denote all agent alignments and the notation $\Phi_{i,j}$ to denote the alignment that agent i has with agent j.

Ontology alignment is only used when communication cannot be understood. This creates the *focus* of the ontology alignment problem. This focus is influenced by pre-defined communication rules, that dictate what agents attempt to communicate about, fixed action policies that influence the 'behaviour' of agents within the environment and a fixed alignment that is a subset of all agents' alignments that specifies a shared alignment between all agents prior to language games. As we are only focused on instance matching, this fixed alignment contains correct equivalences between all non-instance ontological components.

For the purpose of evaluation, we generate gold standard alignments between all agents Φ^*, where $\Phi^*_{i,j}$ is the gold standard alignment between agent i and j. This is created by keeping track of the labels that each agent assigns to $\mathcal{E}$, allowing for a gold standard alignment to be constructed directly from ground truth. Given this gold standard alignment, we can then define our ontology alignment problem as follows: given agents i and j, and their ontologies O and O' respectively, find an alignment $\phi \in \Phi_{i,j}$ from O to O' such that $\phi = \Phi^*_{i,j}$.

4   Proposed Solution Method

Our solution to this ontology alignment problem is based on using 'language games' defined as follows:

Definition 4.1. *A language game is a coordinated communication process between two agents i and j, called the comprehender and the explicator respectively, that generates correspondences between $\mathcal{L}_i$ and $\mathcal{L}_j$. A language game consists of three steps:*

Target Selection: *The comprehender chooses some target label $l_{target} \in \mathcal{L}_j$ that they cannot understand and which they were exposed to in previous communication.*

Context Selection: *The explicator provides 'context' for l_{target} to distinguish it from other labels.*

Correspondence Induction: *Given this context, the comprehender infers a correspondence, $\langle l_{local}, l_{target}, \theta \rangle$, by induction, where θ is a semantic relation that holds between l_{local} and l_{target} from a set Θ of possible semantic relations.*

We consider each step of a language game to be a *strategy*. In this paper we use a fixed strategy for target selection and focus on context selection and correspondence induction. Agents communicate messages to each other, where the content of a message is a graph pattern (described later in this section). A language game follows a simple protocol that is shown in Fig. 2 that dictates the state of communication. This protocol distinguishes between two stages of communication: operational communication, that is communication between agents that is intended to be understood and explication communication, that consists of communication to align ontologies and facilitate future operational communication.

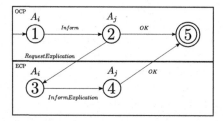

Fig. 2. *Operational communication protocol* (OCP): Agent i sends an *Inform* message to inform agent j about changes in $\mathcal{E}$. If Agent j cannot correctly translate this *Inform* message, agent j requests explication as a *RequestExplication* message. *Explication communication protocol* (ECP): Agent i replies with an *InformExplication* message containing the explication from their content selection strategy. This is a simplified version of the protocol suggested by [13].

All context selection strategies that we consider consist of the *neighbourhood* of the element that l_{target} refers to. This neighbourhood consists of any element from $\mathcal{I} \cup \mathcal{C} \cup \mathcal{T} \cup \mathcal{D}$ that can be reached by traversing from the element corresponding to l_{target} along relations defined by $\mathfrak{R}$. The relations that are traversed along are also included in the neighbourhood, preserving structural information. We restrict the traversed relations to not include *MemberOf* or *HasType* relationships. Without this restriction, the neighbourhood of an instance or data value along two traversed relations could be all instances or data values that share a concept or data type with l_{target}. After the neighbourhood is selected, *MemberOf* and *HasType* relations and their concepts and data types are included. Our assumption is that structural similarity/dissimilarity of neighbourhoods serve as evidence for similarity/dissimilarity between elements of those neighbourhoods, in particular, the target element that corresponds to the target label selected in the target selection strategy.

Though this is a very general structure-based similarity measure, in this paper, we are only concerned with applying it to instance matching. As such, the set of semantic relations Θ only consists of equivalence ($=$), that indicates that l_{local} and l_{target} refer to the same instance in the environment, and disjunction ($\perp$), that indicates that l_{local} and l_{target} do not refer to the same instance in the environment.

To facilitate reasoning about structural similarity, we represent ontologies as vertex and edge labelled directed multi-graphs as follows:

Definition 4.2. *A vertex and edge labelled directed multi-graph is a tuple $G = (V, E, \Sigma_V, \Sigma_E, \ell, s, t)$ where V is a set of vertices, E is a multiset of ordered pairs from $V \times V$, Σ_V is a set of labels for vertices, Σ_E is the set of labels for edges, ℓ is a labelling function $\ell : V \cup E \rightarrow \Sigma_V \cup \Sigma_E$; $s : E \rightarrow V$ assigns each edge to its source vertex; and $t : E \rightarrow V$ assigns each edge to its target.*

We represent agent i's ontology O_i as a graph knowledge base $\mathcal{K}_i$ where $V(\mathcal{K}_i)$ is a set of vertices corresponding to elements of $\mathcal{C} \cup \mathcal{I} \cup \mathcal{T} \cup \mathcal{D}$, $E(\mathcal{K}_i)$ is a set of edges derived from $\mathfrak{R}$ and t and s are defined such that they respect the ordering of pairs in $E(\mathcal{K}_i)$, i.e., $t(\langle v, v' \rangle) = v'$ and $s(\langle v, v' \rangle) = v$.

Agents communicate about their local knowledge by exchanging 'graph patterns' as defined in [1]:

Definition 4.3. *A graph pattern π is a tuple with the same elements as those in Definition 4.2 except $V = V_{const} \cup V_{var}, E = E_{const} \cup E_{var}, \Sigma_V = REG(\Sigma_{V_{const}} \cup \Sigma_{V_{var}}), \Sigma_E = REG(\Sigma_{E_{const}} \cup \Sigma_{E_{var}})$, indicating that vertices and edges can represent either constants or variables and a regular language over vertex and edge labels denoted by $REG(\Gamma)$ which denotes the set of non-empty regular languages over Γ. We denote π_Σ as the graph pattern labelled with $\Sigma_V \cup \Sigma_E$.*

An example of a graph pattern is shown in Fig. 3.

Further to the notion of a graph pattern, is a graph pattern query. This is a pair $Q = (\pi, \bar{x})$ where π is a graph pattern and $\bar{x}$ is a tuple of elements from $V(\pi)$. This is similar to a conjunctive query where $\bar{x}$ is the head of the query, containing distinguished variables and π is the body of the query, limiting what $\bar{x}$ can be bound to. Given a knowledge base $\mathcal{K}$ and a graph query $Q = (\pi, \bar{x})$ with $|\bar{x}| = k$, the answer to Q on $\mathcal{K}$ is:

$$Q(\mathcal{K}) = \{\bar{v} \in V^k | \mathcal{K} \models \pi[\bar{v}/\bar{x}]\}. \tag{3}$$

Here $\pi[\bar{v}/\bar{x}]$ is the result of substituting $\bar{v}$ for $\bar{x}$ in the pattern π. $\bar{x}$ can consist of any vertices in π that are constants or variables, while its substitution

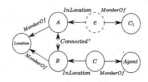

Fig. 3. π_Σ expresses that there is an agent instance labelled 'C' that is within one or more connected relations of a Location labelled either 'A' or 'B' that has some instance denoted by variable 'e' of type C_1 in that location.

$\bar{v}$ consists of constants from $\mathcal{K}$ constrained by the graph structure in π. We refer to π as the *context* for $\bar{x}$, as π serves to distinguish vertices in $\bar{x}$ from other vertices by constraining how it can map onto a knowledge base.

In our language games, the explicator provides context for l_{target} as a graph pattern query where $\bar{x}$ is a single vertex corresponding to l_{target} and π contextualises l_{target}. The comprehender matches this context against their local knowledge base, finding possible valuations for $\bar{x}$ and hence also for l_{target}. When $|Q(\mathcal{K})| > 1$ the answer to the query, and hence the context provided by the explicator, is ambiguous. The higher the cardinality of $Q(\mathcal{K})$, the more ambiguous the context is. An unambiguous graph query is then one where $|Q(\mathcal{K})| - 1$. It is also possible that $|Q(\mathcal{K})| = 0$, indicating that the context provided by the explicator does not overlap with the comprehender's knowledge. This is expected to occur given that agents have heterogeneous knowledge. Reasoning about ambiguity features prominently in our language game strategies described later.

Before the comprehender matches context from the explicator, the comprehender first translates the context according to a mapping function between sets of labels,

$$map_{i,j}: \mathcal{L}_j \nrightarrow \mathcal{L}_i \qquad (4)$$

where $map_{i,j}$ is agent i's mapping function for agent j such that $map_{i,j}(l') = l \iff \langle l, l', = \rangle \in \Phi_{i,j}$ where $l' \in \mathcal{L}_j$, $l \in \mathcal{L}_i$. $map_{i,j}$ is a partial function as agent i does not have a complete mapping from agent j's labels to their own in general. A graph query is then translated as follows: if a mapping for a label belonging to a constant vertex is defined, this label is substituted by its mapping. Otherwise, the vertex corresponding to the constant label is moved to V_{var} making it a variable. The label given to this vertex depends on whether there are known disjunctions or not for the constant label according to alignment $\Phi_{i,j}$. If there are no known disjunctions, the label is given a unique variable label. If there are known disjunction, a regular expression of possible alternatives for the constant is created indicating that the label could be one of any label of the same concept for which a disjunction semantic relation does not hold. For example, a translation for a vertex might be $A|B|C$, indicating that the vertex label is either A or B or C.

An example of the graph matching that occurs in a language game is given in Fig. 4.

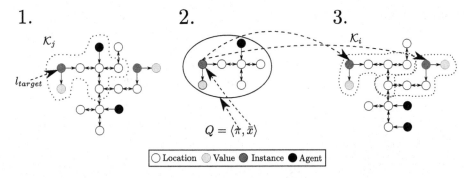

Fig. 4. Graph matching in a language game. For clarity, these graphs omit vertices coresponding to members of $\mathcal{C}$ and $\mathcal{T}$ as well as *MemberOf* and *HasType* relations. 1. The explicator selects the neighbourhood of a vertex corresponding to l_{target}. 2. This neighbourhood is communicated to the comprehender as a graph query. At this point, the graph query is translated by the comprehender according to their current mapping and alignment with the explicator. 3. The comprehender finds all valuations for this graph query against their knowledge base. In this case, there is no exact match for the context given by the explicator. However, there are two partial matches that provide evidence that l_{target} may be one of two vertices. The partial matches are shown by dotted boundaries and the potential candidates for l_{target} are indicated by dashed arrows from the query to $\mathcal{K}_i$.

Reasoning About Environmental Dynamics. To reason about environmental dynamics, agents must maintain a model of 'change' of their environment. An edge has 'changed' with respect to local knowledge if: there is evidence that an observed edge (those edges that they have observed in previous time steps) present in the knowledge base in the last time step, is no longer present, or, if an edge that was not present in the knowledge base in the previous time step is now present. An edge is 'unchanged' if it existed in the knowledge base in the previous time step and still exists in the knowledge base. Each agent maintains a data set for each edge they have observed. This data set is a set of tuples of the form $\langle x, y \rangle$, where y is a label indicating 'changed' or 'unchanged' and x is the elapsed time since the last observed change. For example, $\langle 5, changed \rangle$ for an observed edge indicates that the edge had changed after 5 time steps. This serves as a training set for a support vector machine [4] which is used in Sect. 4.1 to classify whether an edge has changed or not since it was last observed. We use this notion of change and this classification technique whenever updating knowledge and reasoning about uncertain local knowledge respectively.

4.1 Language Game Strategies

In this section we begin by formalising language game strategies in general and followed this by describing particular strategies used in our experiments. We omit a general description of 'target selection' as this is hard-coded in our experiments. Correspondence induction and context selection are defined as follows:

Definition 4.4. *Correspondence induction. Given the comprehender's knowledge base $\mathcal{K}_i$, l_{target}, context Q from the explicator such that l_{target} is the label of a vertex in $V(\pi)$ where $\pi \in Q$, and the existing alignment of the comprehender with the explicator ϕ, induce a set of correspondences $\phi' \subseteq \mathcal{L}_i \times l_{target} \times \Theta$ if Q provides enough evidence that $\phi' \subseteq \Phi^*_{i,j}$ and update ϕ with these.*

Definition 4.5. *Context selection. Given the explicator's knowledge base $\mathcal{K}_j$, $l_{target} \in \mathcal{L}_j$, the existing alignment of the explicator with the comprehender ϕ and a correspondence induction strategy, send a message to the comprehender with a graph pattern Q such that l_{target} is in $\bar{x}$ where $\bar{x} \in Q$ and such that the comprehender, using the correspondence induction strategy with Q, will induce correct correspondences.*

The difficulty of these sub-problems stem from the distributed nature of these language games. Agents do not have access to each other's knowledge and even if they did, their knowledge is labelled differently. As such, any solution to this problem relies on the assumption that structural similarity between a translated query and a knowledge base is enough evidence for induction of correct correspondences. Also note that, though we refer to the gold standard alignment Φ^* in our definitions, this is of course inaccessible to agents when reasoning.

Target Selection Strategy. When agent i receives an operational message from agent j that cannot be translated (i.e. $map_{i,j}$ is undefined for any label in the message), agent i sends an ExplicationRequest message to agent j for the labels that have an undefined mapping. Agent i then discards the operational message. Labels are no longer selected for target selection if the explication by the explicator results in no matches ($Q(\mathcal{K}_i) = 0$).

Correspondence Induction Strategies. The following correspondence induction strategies take as input the parameters described in Definition 4.4. The graph queries in correspondence induction are first translated before serving as input for these strategies.

Exact Match: Correspondences are induced as follows: if $|Q(\mathcal{K})| = 1$, induce $\langle \ell(v)$ where $v \in Q(\mathcal{K})$, l_{target}, $=\rangle$; if$|Q(\mathcal{K})| > 1$ induce $\{\langle \ell(v), l_{target}, \perp \rangle | v \notin Q(\mathcal{K}) \wedge v \in V(\mathcal{K})\}$ indicating that l_{target} is disjoint ($\perp$) from some local labels. This is essentially induction based on graph isomorphism where labels, regular expressions and graph structure must match.

Partial Match: The maximum common sub graph containing a vertex for l_{target} between π and $\mathcal{K}$. The consequence of this is that edges in π become optional and so can be removed from the query if they do not match. If there are multiple maximum common sub graphs that bind l_{target} to the same vertex, only one of these is selected at random for each possible binding of l_{target}. Induction is then handled in the same way as in the exact match strategy.

Context Selection Strategies. The following strategies take as input the parameters described in Definition 4.5.

K-Increasing: We define an algorithm $kcon$ that, given a vertex $v \in V(\mathcal{K})$ and a natural number k, returns the neighbourhood within k edges of v. The k-increasing strategy begins with $k = 0$ and generates a query $Q = \langle v, v \rangle$ where of course v is the only sub graph within 0 edges of v. k is then increased by 1 for each subsequent request of v. The value k associated with requests for explication of a vertex is independent between vertices and requests from different agents. This essentially expands the neighbourhood of vertex v where $l_{target} = \ell(v)$ each time a request is made.

K-Selection the explicator chooses the size of the neighbourhood to share by applying the k-increasing strategy against their own knowledge base. The neighbourhood selected is the lowest value of k that is locally unambiguous (when $|Q(\mathcal{K})| = 1$). Before executing the query, the context selected by the k-increasing strategy is translated in reverse: only labels that the explicator believes the comprehender understands are included in the context. Intuitively this answers the hypothetical question: what context would the explicator need to provide to their self to disambiguate a label?

Uncertainty Removal: We consider variations of k-increasing and k-selection where uncertain edges are removed from graph patterns. This is applied to the graph selected by $kcon$ so that k-selection selects only certain context. We also explore an extreme version of uncertainty removal where all dynamic edges are removed, leaving only static edges.

5   Experimentation

The environment is generated pseudo-randomly: there is a randomly generated component and some predefined relational structure. The environment we use for experimentation is a grid world, where cells are locations, there are agents and instances in cells and there are arbitrarily many connections between locations. The generation of $\mathcal{E}$ is parametrised by a vector $\mathbb{N}^m$ where each parameter is used to specifying the number of vertices in the generated graph. This vector corresponds to $\langle locations, agents, otherInstances, otherConcepts \rangle$. This allows us to easily specify randomly generated ontologies with only a few parameters. The ontology we consider is:

$$O = \langle \mathcal{C} = \{Location, Agent, C_1, C_2, \cdots, C_{\mathbf{E}_4}\},$$
$$\mathcal{I} = \{I_1, I_2, \cdots I_{\mathbf{E}_1 + \mathbf{E}_2 + \mathbf{E}_3 \cdot \mathbf{E}_4}\}, \mathcal{T} = \{Boolean\},$$
$$\mathcal{D} = \{D_1, D_2, \cdots, D_{\mathbf{E}_3 \cdot \mathbf{E}_4}\},$$
$$\mathcal{R} = \{Connected, InLocation, MemberOf, HasType, HasValue\}, \mathfrak{R} \rangle$$

The instances $I_{\mathbf{E}_1 + \mathbf{E}_2 + 1}, I_{\mathbf{E}_1 + \mathbf{E}_2 + 2}, \cdots, I_{\mathbf{E}_1 + \mathbf{E}_2 + \mathbf{E}_3 \cdot \mathbf{E}_4}$, of classes $C_1, C_2, \cdots, C_{\mathbf{E}_4}$, are the possible target instances of our language games. Data values in $\mathcal{D}$ are properties of these instances where: data values are not shared between instances. There are initially an even number of true and false instances.

$MemberOf$ relations are created such that $\forall i \in \mathcal{I}$, $\exists c \in \mathcal{C}$ s.t $\mathfrak{R}(i, c) = \{MemberOf\}$. In particular:

- $\forall location \in \{\mathcal{I}_1, \mathcal{I}_2, \cdots, \mathcal{I}_{\mathbf{E}_1}\}$, $\mathfrak{R}(location, Location) = \{MemberOf\}$.
- $\forall agent \in \{\mathcal{I}_{\mathbf{E}_1+1}, \mathcal{I}_{\mathbf{E}_1+2}, \cdots, \mathcal{I}_{\mathbf{E}_1+\mathbf{E}_2}\}$ $\mathfrak{R}(agent, Agent) = \{MemberOf\}$.
- Each target instance has a $MemberOf$ relation with a single concept from $C_1, C_2, \cdots, C_{\mathbf{E}_4}$ such that there are an even number of instances per concept.

$HasType$ relations are created such that all elements of $\mathcal{D}$ have the type $Boolean$. $InLocation$ relations are created incrementally between non-location instances and a randomly selected location that does not already contain an instance of the same type. If two instances of the same type are in the same location, they would not be distinguishable from one another.

$Connected$ relations are randomly generated between $Location$ instances using a variation of Erdős-Rényi $G(n, p)$ [9] random graph model where vertices (n) are $Location$ instances and edges are $Connected$ relations that hold from one location to any other location with a probability of p. To ensure that locations are fully connected, we connect a random vertex in one component to a random vertex in another component until the graph is fully connected. An example of a generated environment is shown in Fig. 1. This is generated with parameters $\langle 2, 2, 1, 1 \rangle$.

5.1   Environment Dynamics

Only data values and the location of agents change in the environment. When the environment is created, there are initially an even number of $True$ and $False$ data values. Values $d \in \mathcal{D}$ then alternate between $True$ and $False$ according to a parameter $\rho \in (0, 1]$ where each $d \in \mathcal{D}$ alternates value with a probability ρ at each time step. Agents create plans to move stochastically in the environment. When agent i does not have a plan, it selects a location at random to travel to. It uses A* search to plan a path along $Connected$ relations from its current location to its target location. At each time step it moves along one $Connected$ relation into a new location. When agent i arrives at its target location, agent i re-plans in the same way.

5.2   Measurements

Measurement $\hat{\mathcal{E}} = \mu_i^t(\mathcal{E})$ is received by agent i at time t. The set of vertices received by agent i are: A vertex representing agent i, the location agent i is in, any instances in agent i's location (including other agents), any data values and data types of these instances, all locations along one $Connected$ relation from their current location and all instances in these locations but not their properties. For example, in Fig. 1, agent D would measure them self, locations B and C, agent B and instance A as well as all $Connected$, $InLocation$ and $MemberOf$ relations. Agent E would measure the entire graph.

5.3   Operational Communication

Agents can communicate with each other from any location and at any time. Operational communication in this scenario consists of messages indicating that an instance has changed data value. When an agent observes this change, the agent communicates the instance and its new data value. The purpose of this communication is to enable agents to have more complete knowledge of the environment by sharing their measurements.

5.4   Updating $\mathcal{K}$

At the start of each experiment, agents are given a snapshot of the complete environment. Agents duplicate this snapshot as their local knowledge. Agents then spend a fixed amount of time learning the dynamics of the environment from measurements. During this time, differences in measurements and behaviour result in different knowledge.

Agents use common sense rules to detect change to update their knowledge: 1. An agent cannot be in two locations at once, therefore if an agent is seen in a new location, it no longer holds that it is in the old location. 2. There can be only one data value that is a property of an instance with a particular relation name, therefore a new value for this property overwrites an old value. Given these definitions of change, the way in which agents update their knowledge from successfully translated operational messages and measurements is the same: new edges are added if they do not occur in $\mathcal{K}$, and inconsistent edges are removed.

6   Experimental Results

We compare pairs of context selection and correspondence induction strategies with respect to correctness of alignments and the amount of context required to achieve this. To measure the amount of context sent, we count the number of edges sent in explication messages, excluding edges that indicate class membership, i.e., *HasType* edges. We then average the number of edges across the number of explication messages sent.

To measure correctness of alignments, we use semantic precision and recall described by [6]. Given an alignment $\phi \in \Phi_{i,j}$ and the gold standard reference alignment $\phi^* \in \Phi_{i,j}^*$, semantic precision and recall is calculated as $P(\phi, \phi*) = \frac{C(\phi) \cap C(\phi*)}{C(\phi)}$ and $R(\phi, \phi*) = \frac{C(\phi) \cap C(\phi*)}{C(\phi*)}$ where $C(\cdot)$ is the deductive closure of an alignment under its entailments, i.e., all alignments that can be deduced from other alignments. We use *F-score* as a combined measurement of precision and recall defined as $F\text{-}score(\phi, \phi*) = 2 \cdot \frac{P(\phi,\phi*) \cdot R(\phi,\phi*)}{P(\phi,\phi*) + R(\phi,\phi*)}$. We then use the mean *F-score* across all alignments of all agents excluding the fixed known alignments. Semantic precision and recall is often not possible to compute in the general case. However, our entailments are simple: If an equivalence ($=$) between two labels that refer to instances of the same class is known, a disjoint ($\perp$) semantic relation

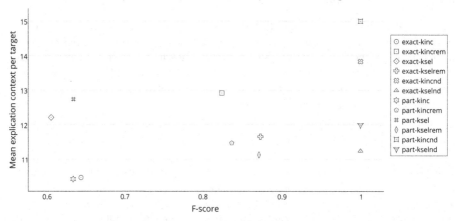

Fig. 5. *F-score* and explication content communicated for pairs of correspondence induction and context selection strategies. Legends abbreviations: exact = exact matching, part = partial matching, kinc = k-increasing, kincrem = k-increasing with uncertainty removal, ksel = k-selection, kselrem = k-selection with uncertainty removal, kselnd = k-selection with no dynamics, kincnd = k-increasing with no dynamics.

is deduced between these two labels and all other labels of instances of the same class.

The results of our experiments are shown in Fig. 5. The parameters used in all experiments are : $\rho = 0.25$, $\langle 50, 4, 4, 3 \rangle$ as environment generation parameters, and $n = 50$, $p = 0.1$ as the random graph parameters. Agents learn dynamics of their environment for 1000 steps before communicating and all experiments are run over 10 repetitions. A repetition is complete when all agents have found correct or incorrect equivalence correspondences for all instances, or when they have exhausted attempts at finding correspondences. For the later case we set a limit of 10 requests for explication for an instance target.

Our results show that the removal of uncertain context results in higher F-scores. In the case that agents only communicate static context, agents achieve optimal *F-scores*. This demonstrates that the environment is simple enough, and the agent strategies powerful enough, to resolve the ontology alignment problem by the communication of only static context. Moreover, the amount of context provided to do so is quite modest; only 11 edges per target for the best performing strategy pair (exact-kselnd).

We expected inclusion of some dynamic context to perform better than communicating static context. For example, if agents are in the same location and attempt to communicate about a target in that location, context that includes the agents in the description should result in a correct correspondence with less context than omitting the agents from the description. This suggests that agents' knowledge of dynamic elements of the environment are still too different to be used successfully as context under our uncertainty removal approach.

K-selection strategies result in lower F-scores than k-increasing strategies when the selected context is unsuitable for comprehension by the comprehender. This is because parameter k in the k-increasing strategy is essentially fine-tuned by the comprehension need of the comprehender via repeated requests for explication while the k-selection strategy is blind to the comprehension requirements of the comprehender. However, when the assumption about what context is needed is correct, the same *F-score* can be achieved with much less context (exact-kselnd vs. exact-kincnd). The best performing strategies in our results achieve a perfect *F-score* with and average of between 11 and 15 edges of context per target node. This is quite a modest communication overhead requirement to enable correct interpretation of messages.

7   Discussion

The context selection strategies that we have explored in this paper focus on finding context that is both unambiguous and shared between agents. In future work, we plan to extend context selection strategies by both identifying and exploiting salient features of the environment and including approximate dynamic information in selected context. One can identify salient parts of the environment by statistical analysis of graphs and use this information to bias context selection strategies towards more salient parts of the environment. Our results have shown that simply removing uncertain dynamic context only goes so far in improving alignment success. Rather than removing uncertain context all together, bounding uncertainty by providing approximate context may be beneficial. For example, if the location of an agent is uncertain, rather than excluding the agent from explication, the explicator can include the agent's approximate location as a regular expression over *Connected* relations bound by their maximum expected distance travelled since they were last observed.

As well as expanding on context selection strategies, we plan to explore target selection strategies. In particular, when operational communication messages contain multiple misunderstood labels, or more generally, when an agent has a pool of possible misunderstood labels to select from, the comprehender must choose a sequence of target labels as the focus of language games. Further to this, the comprehender can select multiple targets in a single explication request, requiring that the explicator disambiguates all of these targets with a single explication message.

8   Conclusion

In this paper, we proposed a novel combination of language games and graph-based knowledge representation as a solution to decentralised ontology matching between agents situated in a shared environment where ontologies are representations of agents' beliefs about their environment. To this end, we defined a language game as a sequence of three strategies: target selection, correspondence induction and context selection. We compared the performance of various correspondence induction and context selection strategies given a fixed target selection strategy.

Our results show that structural similarity alone can help align ontologies that are at the same level of granularity without agents utilising grounding through physical interaction with only a modest communication overhead. However, environmental dynamics and incomplete measurements that result in different local knowledge must be reasoned about for this to be possible. We have also shown that the shortcomings of a correspondence induction strategy can be ameliorated by the choice of context selection strategy and vice versa.

In future work we plan to explore more complex language game strategies. In particular, context selection strategies that identify and reason about salient features of the environment and the inclusion of approximate dynamic information as context, and target selection strategies, where agents must select sequences of targets for language games and where a single language game can involve multiple targets.

Acknowledgments. The research reported in this paper is supported by an EPSRC Industrial CASE Award Ph.D. Studentship with Selex ES Ltd.

References

1. Barceló, P., Libkin, L., Reutter, J.L.: Querying graph patterns. In: Proceedings of the Thirtieth ACM SIGMOD-SIGACT-SIGART Symposium on Principles of Database Systems, pp. 199–210. ACM (2011)
2. Cangelosi, A.: The grounding and sharing of symbols. Pragmatics Cogn. **14**(2), 275–285 (2006)
3. Coradeschi, S., Saffiotti, A.: An introduction to the anchoring problem. Robot. Auton. Syst. **43**(2), 85–96 (2003)
4. Cortes, C., Vapnik, V.: Support-vector networks. Mach. Learn. **20**(3), 273–297 (1995)
5. Davidovsky, M., Ermolayev, V., Tolok, V.: A survey on agent-based ontology alignment. In: ICAART, vol. 2, pp. 355–361. Citeseer (2012)
6. Euzenat, J.: Semantic precision and recall for ontology alignment evaluation. In: IJCAI, pp. 348–353 (2007)
7. Euzenat, J., Shvaiko, P.: Ontology Matching, 2nd edn. Springer, Heidelberg (2013)
8. Fensel, D., McGuiness, D.L., Schulten, E., Ng, W.K., Lim, G.P., Yan, G.: Ontologies and electronic commerce. IEEE Intel. Syst. **16**(1), 8–14 (2001)
9. Gilbert, E.N.: Random graphs. Ann. Math. Stat. **30**(4), 1141–1144 (1959)
10. Harnad, S.: The symbol grounding problem. Phys. D: Nonlinear Phenom. **42**(1), 335–346 (1990)
11. McNeill, F., Bundy, A.: Dynamic, automatic, first-order ontology repair by diagnosis of failed plan execution. Int. J. Seman. Web Inf. Syst. (IJSWIS) **3**(3), 1–35 (2007)
12. Steels, L.: The Talking Heads Experiment. Volume 1. Words and Meanings. Laboratorium, Antwerpen (1999)

13. van Diggelen, J., de Jong, E.D., Wiering, M.A.: Strategies for ontology negotiation: finding the right level of generality. In: Dignum, F.P.M., van Eijk, R.M., Flores, R. (eds.) AC 2005. LNCS (LNAI), vol. 3859, pp. 164–180. Springer, Heidelberg (2006)
14. Vogt, P.: The physical symbol grounding problem. Cogn. Syst. Res. **3**(3), 429–457 (2002)

An Overview of Argumentation Frameworks for Decision Support

Lucas Carstens, Xiuyi Fan, Yang Gao, and Francesca Toni$^{(\boxtimes)}$

Department of Computing, Imperial College London, London, UK
{lucas.carstens10,x.fan09,y.gao11,f.toni}@imperial.ac.uk

Abstract. Several forms of argumentation frameworks have been used to support decision-making: these frameworks allow at the same time a graphical representation of decision problems as well as an automatic evaluation of the goodness of decisions. We overview several such uses of argumentation frameworks and discuss future directions of research, including cross-fertilisations amongst them.

1 Introduction

Argumentation frameworks (AFs), a non-monotonic reasoning paradigm that consists of a set of arguments and relations between these arguments, have attracted considerable research attention in recent years [28]. AFs can be naturally represented as directed graphs, with each node representing an argument and each arc representing an attack or, in certain AF definitions, a support. By using a graphical representation of AFs, one can not only represent domain knowledge and conflicts therein, but can also perform reasoning over this knowledge in an intuitive manner. These properties make AFs particularly suitable for supporting decision making problems [1], as, in addition to affording a graphical view of decision problems, argumentation provides the capability to evaluate graphs, with the aim of reaching 'dialectically justified' decisions. In this paper, we provide an overview of a number of AFs and applications thereof offering various ways of supporting some form of decision-making. We also discuss promising directions for future research in this area.

We consider multiple forms of AFs, from classic *Abstract AFs* [11] to their extensions, namely *bipolar AFs* (BAFs) [8] and *Value-Based AFs* (VAFs) [5], to structured argumentation frameworks in the form of *Assumption-Based AFs* [13,30], referred to as ABA frameworks. From the application domain perspective, we consider decision-making based on *Question & Answer* (Q&A), *collaborative multi-agent* decision making and classic *multi-criteria* decision making. We describe an application domain for each form of argumentation to illustrate their concrete use, and explain why the chosen form of argumentation is suitable for each particular problem.

One may naturally opt for different combinations of AFs and application domains than those described in this paper. We discuss future directions and the possibility and advantages of such cross-fertilisation towards the end of this paper.

© Springer International Publishing Switzerland 2015
M. Croitoru et al. (Eds.): GKR 2015, LNAI 9501, pp. 32–49, 2015.
DOI: 10.1007/978-3-319-28702-7_3

A1 ⟵————————⟶ A2   A1 ⟵———————— A2
(a) AF (Section 2)   (b) AF⁻ (Section 4.1)

Fig. 1. Example argumentation frameworks.

The remainder of this paper is organised as follows: In Sect. 2, we provide background of AFs; in Sect. 3, we introduce BAFs and outline their use in some Q&A decision making problems; in Sect. 4, we introduce VAFs and describe their use in collaborative multi-agent decision making problems; in Sect. 5, we introduce ABA frameworks and their use in multi-criteria decision making; in Sect. 6, we discuss future work and conclude.

2 Abstract Argumentation

An *Abstract AF* [11] – referred to simply as AF – consists of a set of arguments and a binary *attack* relation between arguments, representing conflicts between arguments. Formally, an AF is a pair

$$(Args, Attack)$$

where *Args* is a set of *arguments* and *Attack* $\subseteq$ *Args* × *Args* is a binary relation ((A, B) ∈ *Attack* is read '*A attacks B*'). An AF can be represented as a directed graph, in which each node corresponds to an argument and each directed arc corresponds to an attack. As an example, consider the following arguments

A1: Let's have dinner at home today
A2: Let's have dinner in a restaurant today.

Since these two arguments support incompatible decisions they attack one another. This AF can be represented by the directed graph in Fig. 1(a).

There are multiple criteria for selecting the 'winning' arguments in an AF, and these criteria are known as *semantics* in argumentation [11]. To be specific, semantics of AFs are defined as 'dialectically acceptable' sets of arguments, where each set is known as an *extension*. For example, for an AF $F = (Args, Attack)$, $S \subseteq Args$ is an *admissible* extension iff S is *conflict-free* (i.e. for any argument $A \in S$, A is not attacked by any argument in S) and can *defend* all its member arguments (i.e. for any $A \in S$, if there exists an argument $B \in Args \setminus S$ that attacks A, there is some argument in S attacking B). Still consider the AF in Fig. 1(a). {**A1**}, {**A2**} and $\emptyset$ are all admissible extensions, whereas {**A1, A2**} is not, because it is not conflict-free.

Abstract Argumentation is the most widely used form of argumentation and offers great simplicity. Also, most other forms of argumentation are either instances or extensions of Abstract Argumentation. For example, Assumption-Based Argumentation [13,30] (see also Sect. 5.1) is an instance of Abstract Argumentation and both Bipolar Abstract Argumentation [8] (see also Sect. 3.1) and Value-Based Argumentation [5] (see also Sect. 4.1) are extensions of Abstract Argumentation.

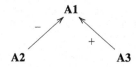

Fig. 2. Example bipolar argumentation framework (− stands for attack, + stands for support, see Sect. 3.1)

3   Bipolar Abstract Argumentation and Q&A-Based Decision-Making

We first give some background in Bipolar Abstract Argumentation (Sect. 3.1) and then outline its use to support a Q&A system (www.quaestio-it.com, Sect. 3.2) and Q&A-based decision-making in engineering design (Sect. 3.3).

3.1   Bipolar Abstract Argumentation

A *Bipolar Abstract Argumentation Framework* (BAF) [8] is an AF extended with a binary *support* relation between arguments. Formally, a BAF is a triple

$$(Args, Attack, Support)$$

where $(Args, Attack)$ is an AF and $Support \subseteq Args \times Args$ is a binary relation ($(A, B) \in Support$ is read 'A supports B'). A BAF can be represented as a directed graph, in which each node corresponds to an argument and each directed arc corresponds to an attack or a support (thus arcs need to be labelled accordingly). Take, for example, the below excerpt of a discussion between John, Joe and Jane on whether or not they should go and watch the latest *Avengers* movie in the cinema:

> John: I think we should go and see the new Avengers; the first one was really great! (**A1**)
> Joe: Please spare me! It's just going to be another big Hollywood production that goes for explosions instead of plot and characters. (**A2**)
> Jane: I loved the first one, as well, so I think we should see it! (**A3**)

By identifying that Joe disagrees with (attacks) John and Jane agrees with (supports) John, this dialogue can be mapped onto the BAF shown in Fig. 2.

There are multiple criteria for selecting the 'winning' arguments in a BAF, and, as in the case of other AFs, these criteria are known as *semantics* [8]. Some of these semantics are defined as 'rationally acceptable' extensions, analogously to Abstract AFs. Here, however, we focus on a class of *quantitative semantics*, assessing the 'dialectical' strength of arguments numerically. In particular, we focus on two of these semantics, given in [3,4,19] respectively, both building upon [24]. These approaches are referred to as QuAD (for Quantitative

Argumentation Debate) [3,4] and ESAA (for Extended Social Abstract Argumentation) [19]. Both QuAD and ESAA assume that arguments are equipped with a *base score*, namely a number in the $[0,1]$ interval. ESAA also assume that positive and negative votes may be ascribed to arguments, that result in a modification of their base score (see [19] for details). In both approaches, the (given or modified) base score amounts to an intrinsic (non-dialectical) strength of arguments. Both approaches determine the (dialectical) strength of arguments by aggregating the strength of attackers against and supporters for these arguments, for restricted types of BAFs in the form of trees. Let v_0 be the base score of an argument A, and v_a, v_s the combined strength of all attackers against A and supporters for A, respectively. Then the dialectical strength of A is given by $g(v_0, v_a, v_s)$ defined as follows:

$$g(v_0, v_a, v_s) = v_a \text{ if } v_s = nil \text{ and } v_a \neq nil$$
$$g(v_0, v_a, v_s) = v_s \text{ if } v_a = nil \text{ and } v_s \neq nil$$
$$g(v_0, v_a, v_s) = v_0 \text{ if } v_a = v_s = nil$$
$$g(v_0, v_a, v_s) = \frac{(v_a + v_s)}{2} \text{ otherwise}$$

Here $v_a = nil/v_s = nil$ if there are no attackers against/supporters for (respectively) A or all such attackers/supporters have strength 0 (and are thus ineffective [4]). The combined strength v_a/v_s of (the sequence S of the strength of) all attackers against/supporters for A is computed as $\mathcal{F}_*(v_0, S)$, for $*$ a or s (respectively), defined recursively as follows, in both QuAD and ESAA:

$$\text{if } S \text{ is ineffective} : v_* = nil$$
$$\text{if } S = (v) : \mathcal{F}_*(v_0, S) = f_*(v_0, v)$$
$$\text{if } S = (v_1, \ldots, v_n) : \mathcal{F}_*(v_0, (v_1, \ldots, v_n)) =$$
$$f_*(\mathcal{F}_*(v_0, (v_1, \ldots, v_{n-1})), v_n)$$

where in both QuAD and ESAA:

$$f_a(v_0, v) = v_0 - v_0 \cdot v = v_0 \cdot (1 - v)$$

in QuAD:

$$f_s(v_0, v) = v_0 + (1 - v_0) \cdot v = v_0 + v - v_0 \cdot v$$

and in ESAA:[1]

$$f_s(v_0, v) = min(v_0 + (v_0 - v_0 \cdot (1 - v)), 1)$$

---

[1] This presentation of f_* corrects a typo in [19]. There, the *min* condition was erroneously omitted.

For example, given the BAF of Fig. 2, assuming that all arguments have a base score of 0.5, the dialectical strength of **A1** is given, in both QuAD and ESAA, by

$$g(0.5, f_a(0.5, v_{A2}), f_s(0.5, v_{A3}))$$

where v_{A2}/v_{A3} is the dialectical strength of **A2**/**A3**, respectively. Since there are no attacks against or supports for **A2** and **A3**, there strength is their base score (the third case in the definition of g applies). Thus, in both approaches, the dialectical strength of **A1** is given by

$$g(0.5, f_a(0.5, 0.5), f_s(0.5, 0.5))$$

In both approaches $f_a(0.5, 0.5) = 0.5 \cdot (1 - 0.5) = 0.25$. In QuAD, $f_s(0.5, 0.5) = 0.5 + (1 - 0.5) \cdot 0.5 = 0.75$. In ESAA, $f_s(0.5, 0.5) = min(0.5 + (0.5 - 0.5 \cdot (1 - 0.5)), 1) = min(0.75, 1) = 0.75$. Thus, in both approaches, the dialectical strength of **A1** is

$$(0.25 + 0.75)/2 = 0.5.$$

(It is easy to see however that the two approaches give different strengths in general, as for example discussed in [4].)

3.2   Quaestio-It for Q&A

Quaestio-it (www.quaestio-it.com) is a web-based Q&A debating platform that allows users to open topics, ask questions, post answers, debate and vote. Figure 3 shows the visualisation of a debate aiming at identifying the best material to use for the production of a rubber item (this is an example of a debate in a design engineering domain, see Sect. 3.3). In the figure, the answer (reply) on the right is being debated: it is supported by the comment on the bottom right and attacked, on the left, by a comment that is further attacked (on the bottom left).

In Quaestio-it each answer is the root of a tree, which forms a BAF. Through any evaluation algorithm overviewed in Sect. 3.1, the strength of answers and comments, as determined by the opinions in the community, can be computed. Figure 3 uses the ESAA algorithm. As illustrated in the figure, Quaestio-it visualises the strength of opinion graphically: a larger node indicates a higher strength.

In addition to contributing to the debate, users can also propose new answers, as well as cast votes on answers and opinions of others positively or negatively, increasing or decreasing the strength of the argument voted on, as dictated by the ESAA algorithm. Finally, users can signal answers and opinions of others as *spam* (casting a third kind of vote), so as to limit malicious behaviour by users: after a predefined number of spam votes, answers and opinions (as well as any debate about them) are hidden to users.

Compared with other Q&A platforms, Quaestio-it allows the visualisation of debates as graphs, while also providing an evaluation of the positions in the debate that can, arguably, more effectively and directly inform the decision-making processes underlying the debates.

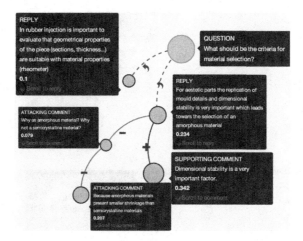

Fig. 3. Visualisation of a debate on a design problem faced by injection molding experts

3.3   Q&A for Engineering Design

The Q&A-style of decision support discussed in the earlier section lends itself well to support decision-making in engineering design, where several stakeholders, with varying degrees of expertise and technical competences, are involved [3,4,6,18]. In particular, this style has been adopted within the DesMOLD project (www.desmold.eu) to facilitate decision making in injection molding. More specifically, the DesMOLD system is an interactive tool that incorporates Quaestio-it to support injection molding experts throughout the design process. The platform is mainly composed of the following processes: (I) a decomposition process to convert complex geometries into simplified geometries, (II) a debate process supporting argumentation and ontology interoperability to ensure designers' mutual understandings and (III) automatic recommendations based on debates and past experience (see [6,18] for details).

As an illustration, Fig. 3 shows a possible debate amongst stakeholders in injection molding. Figure 4 shows how that debate can incorporate contents from past debates, to create a detailed graphical view of a design problem under consideration. Moreover, the evaluation of strength of arguments, as discussed for Quaestio-it in the earlier section, helps identify strong contenders and weak points in the decision problem.

4   Value-Based Argumentation and Collaborative Multi-agent Decisions

We first give some background on Value-based Argumentation (Sect. 4.1) and then outline its use to support Collaborative Multi-Agent decisions (Sect. 4.2).

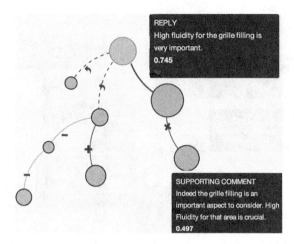

Fig. 4. Visualisation of a debate in which part of a past debate has been inserted, highlighted via a thick blue edge on the right (Color figure online).

4.1 Value-Based Argumentation

Another direction for extending Abstract Argumentation is to consider 'values' promoted by arguments in an argumentation framework. *Value-based AFs* (VAFs) [5] incorporate values and preferences over values into AFs. The key idea is to allow for attacks to succeed or fail, depending on the relative worth of the values promoted by the competing arguments. Given a set V of values, an *audience $Valpref$* is a strict partial order over V (corresponding to the preferences of an agent), and an *audience-specific VAF* is a tuple $(Args, Attack, V, val, Valpref)$, where $(Args, Attack)$ is an AF and $val : Args \rightarrow V$ gives the values promoted by arguments. In VAFs, the ordering over values, $Valpref$, is taken into account in the definition of extensions. The *simplification* of an audience-specific VAF is the AF $(Args, Attack^-)$, where $(A, B) \in Attack^-$ iff $(A, B) \in Attack$ and $val(B)$ is not higher than $val(A)$ in $Valpref$. $(A, B) \in Att^-$ is read 'A defeats B'. Then, (acceptable) extensions of a VAF are defined as (acceptable) extensions of its simplification $(Args, Attack^-)$.

Let us consider the AF introduced earlier in Sect. 2. Let **A1** and **A2** promote values $V1$: Money-saving and $V2$: Time-saving, with $V2$ more preferred than $V1$ in $ValPref$ (denoted as $V2 >_v V1$). Then we can simplify the original AF by eliminating the attack from **A1** to **A2**, because **A2** promotes a higher ranked value than **A1** does. The *simplified argumentation framework* (denoted as AF^-) is illustrated in Fig. 1(b). In AF^-, $\{$**A2**$\}$ is an admissible extension, whereas $\{$**A1**$\}$ is not as it cannot defend its member from the attack from **A2**.

4.2 Collaborative Multi-agent Decisions

VAFs can be used to model domain knowledge to coordinate independent agents in collaborative multi-agent systems (CMAS), where coordination is defined as 'the ability of two or more agents to jointly reach a consensus over which actions to perform in an environment' [23]. VAF-based domain knowledge is useful for specifying heuristics for agents that learn to coordinate efficiently [22].

Consider a scenario of a multi-agent Wumpus World[2] illustrated in Fig. 5. The goal of the agents is to collect all gold and arrive at the exit as quickly as possible, without being killed by a Wumpus. Note that an agent can smell *stench* if it is (non-diagonally) next to a Wumpus, and it can see *glitter* if there is a gold in the agent's square. Suppose we have the domain knowledge for agent Ag1 that there is a Wumpus on its left-hand side, and for agent Ag2 that the exit is next to it on the left-hand side. Also, Ag2 can see that there is a gold in its square. Intuitively, Ag1 should perform action *shoot_left* and Ag2 should perform *pickup*, because there is no need for both agents to shoot the same Wumpus at the same time, and the agents collaboratively need to collect all gold before they exit.

Exit	Ag2 (gold)	Wumpus	Ag1	

Fig. 5. A fragment of a multi-agent Wumpus World game.

To obtain heuristics based on the domain knowledge above using VAFs, first, we can propose four arguments[3]:

- **A1shoot:** Ag1 should perform *shoot_left* because there is a Wumpus next to Ag1 on its left-hand side.
- **A2shoot:** Ag2 should perform *shoot_right* because there is a Wumpus next to Ag2 on its right-hand side.
- **A2left:** Ag2 should perform *go_left* because the exit is on its left-hand side.
- **A2pick:** Ag2 should perform *pickup* because there is a gold in its square.

Then, we can identify conflicts between arguments: for example, arguments **A1shoot** and **A2shoot** conflict, because only one agent needs to shoot the Wumpus. This type of conflict is between different agents' arguments, referred to as *external* conflicts[4]. In addition, arguments **A2shoot** and **A2pick** also conflict, because they recommend the same agent with different actions, while an

---

[2] For ease of presentation, the Wumpus World game used in this paper is slightly different from the classic one presented in [29]. In particular, each agent has nine actions: *go_left, go_right, go_up, go_down, shoot_left, shoot_right, shoot_up, shoot_down* and *pickup*.

[3] Technically, more arguments can be proposed: e.g. an argument 'Ag1 should perform *go_down* to sidestep the Wumpus'. However, for simplicity, we only consider four arguments here.

[4] Two arguments for the same action for two different agents do not necessarily conflict, e.g. the additional argument 'Ag1 should perform *go_left* because it wants to greet the Wumpus' may not conflict with **A2left**, because it is fine that both agents perform *go_left*. Actions that should not be performed by multiple agents can be thought of as *critical*. Thus, external conflicts only exist between arguments recommending the same critical action to different agents.

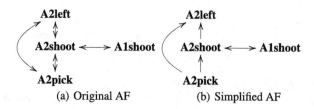

(a) Original AF   (b) Simplified AF

Fig. 6. The original (a) and simplified (b) AF for the scenario in Fig. 5

agent can only perform one action at a time in this setting. This type of conflict, between arguments recommending actions to the same agent, are referred to as *internal* conflicts. Based on these arguments and conflicts, we can build an AF to represent the domain knowledge, as illustrated in Fig. 6(a).

In order to select one best action for each agent, values and preferences over them can be given, as follows:

- *Vsafe:* to make sure the agent is safe.
- *Vmoney:* to make more money.
- *Vexit:* to exit the Wumpus World.

Let **A1shoot** and **A2shoot** promote *Vsafe*, **A2left** promote *Vexit* and **A2pick** promote *Vmoney*. Also, as an example, suppose that $Vmoney >_v Vsafe >_v Vexit$ (for $>_v$, see Sect. 4.1). We can then simplify the AF in Fig. 6(a) by eliminating all attacks pointing to arguments promoting higher ranked values from arguments promoting lower ranked values. The simplified AF (AF^-) is given in Fig. 6(b). We can see that {**A1shoot**, **A2pick**} is the maximal (with respect to $\subseteq$) admissible extension for AF^-, indicating that Ag1 should perform *shoot_left* and Ag2 should perform *pickup*. This can be deemed to be a high-quality heuristic because each agent is recommended one action, and no agent is recommended to perform the same critical action.

Note that VAFs are flexible in that they can easily accommodate changes to CMAS problems. For example, suppose we change the setting of the Wumpus World game, such that the agents only need to arrive at the exit as quickly as possible, without collecting all gold. Given this change, we can simply change the ranking of values to e.g. $Vexit >_v Vsafe >_v Vmoney$ without changing any other components of the VAFs. The new simplified AF is given in Fig. 7. Then, {**A1shoot**, **A2left**} is the maximal (with respect to $\subseteq$) admissible extension, in line with our intuition that, under the new setting, Ag2 should exit instead of picking up gold.

To summarise, in CMAS decision making problems, VAFs not only provide a method for building a graphical representation of (possibly conflicting) domain knowledge, but also provide a mechanism for reasoning over this graphical representation and thus derive heuristics. The examples we provide in this section show that the construction of VAFs can be performed 'semi-automatically' at run-time, 'by instantiating arguments', provided at design-time, in particular

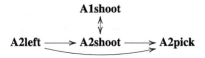

Fig. 7. A new simplified AF of Fig. 6(a) based on the value ranking $Vexit >_v Vsafe >_v Vmoney$.

forms (e.g. each argument recommends an action under certain conditions), values promoted by the arguments and rankings of these values (see [22] for details).

5   Assumption-Based Argumentation and Multi-criteria Decision Making

So far we have focused on variants of Abstract Argumentation, where we abstract away from the content of nodes of argumentation graphs (e.g. sentences in natural language or conditional recommendations for actions) and how edges in these graphs may be determined (e.g. because the sentences agree or disagree or because the recommended actions cannot be executed concurrently). Here we show a further application of argumentation graphs in a form of multi-attribute decision-making, where arguments have a specific structure, in terms of rules and assumptions in an underlying deductive system, and attacks are made on assumptions by deriving their contrary. The form of argumentation we use is Assumption-Based Argumentation (ABA) [13,30]. In Sect. 5.1 we will briefly recap ABA. In Sect. 5.2, we will define a simple (graphical) decision making model and show, in Sect. 5.3, how it can be captured in ABA, benefitting from the graphical representation afforded by ABA.

5.1   Assumption-Based Argumentation (ABA)

ABA frameworks [13,30] are tuples $\langle \mathcal{L}, \mathcal{R}, \mathcal{A}, \mathcal{C} \rangle$ where

- $\langle \mathcal{L}, \mathcal{R} \rangle$ is a deductive system, with $\mathcal{L}$ the *language* and $\mathcal{R}$ a set of *rules* of the form $\beta_0 \leftarrow \beta_1, \ldots, \beta_m (m \geq 0, \beta_i \in \mathcal{L})$;
- $\mathcal{A} \subseteq \mathcal{L}$ is a (non-empty) set, referred to as *assumptions*;
- $\mathcal{C}$ is a total mapping from $\mathcal{A}$ into $2^{\mathcal{L}} \setminus \{\{\}\}$, where each $\beta \in \mathcal{C}(\alpha)$ is a *contrary* of α, for $\alpha \in \mathcal{A}$.

Given a rule ρ of the form $\beta_0 \leftarrow \beta_1, \ldots, \beta_m$, β_0 is referred to as the *head* (denoted $Head(\rho) = \beta_0$) and $\beta_1, \ldots, \beta_m$ as the *body* (denoted $Body(\rho) = \{\beta_1, \ldots, \beta_m\}$). In a *flat* ABA framework assumptions are not heads of rules. Here, we restrict attention to flat ABA frameworks. Indeed, they are sufficient to capture the kinds of multi-attribute decision making problems we are focusing on.

In ABA, *arguments* are deductions of claims using rules and supported by assumptions, and *attacks* are directed at the assumptions in the support of arguments. Informally, following [13,30]:

- *an argument for (the claim)* $\beta \in \mathcal{L}$ *supported by* $\Delta \subseteq \mathcal{A}$ (denoted $\Delta \vdash \beta$) is a finite tree with nodes labelled by sentences in $\mathcal{L}$ or by τ^5, the root labelled by β, leaves either τ or assumptions in Δ, and non-leaves β' with, as children, the elements of the body of some rule with head β'.
- An argument $\Delta_1 \vdash \beta_1$ *attacks* an argument $\Delta_2 \vdash \beta_2$ iff β_1 is a contrary of one of the assumptions in Δ_2.

Thus, for each ABA framework, we can construct a corresponding AF.

Attacks between (sets of) arguments in ABA correspond to attacks between sets of assumptions, where *a set of assumptions* Δ *attacks a set of assumptions* Δ' iff an argument supported by a subset of Δ attacks an argument supported by a subset of Δ'.

Given $\mathcal{F} = \langle \mathcal{L}, \mathcal{R}, \mathcal{A}, \mathcal{C} \rangle$, *a set of assumptions is admissible* (in $\mathcal{F}$) iff it does not attack itself and it attacks all $\Delta \subseteq \mathcal{A}$ that attack it.

We say that an *argument* $\Delta \vdash \beta$ *is admissible (in* $\mathcal{F}$*) supported by* $\Delta' \subseteq \mathcal{A}$ iff $\Delta \subseteq \Delta'$ and Δ' is admissible. We also say that an argument *is in* a framework $\mathcal{F}$ iff all rules used to construct it and all assumptions supporting it are in $\mathcal{F}$.

We will use *Dispute Trees* [12,14] to illustrate how ABA can support explaining best decisions. Given an AF $\langle Args, Attack \rangle$, a *dispute tree for* $A \in Args$ is a (possibly infinite) tree $\mathcal{T}$ such that:

1. every node of $\mathcal{T}$ is labelled by an argument (in the AF) and is assigned the status of either *proponent (P)* or *opponent (O)*, but not both;
2. the root of $\mathcal{T}$ is a P node labelled by A;
3. for every P node n labelled by an argument B, and for every argument C that attacks B (in the AF), there exists a child of n, which is an O node labelled by C;
4. for every O node n labelled by an argument B, there exists at most one child of n which is a P node labelled by an argument which attacks (in the AF) B;
5. there are no other nodes in $\mathcal{T}$ except those given by 1–4.

The set of all arguments labelling P nodes in $\mathcal{T}$ is called the *defence set* of $\mathcal{T}$, denoted by $\mathcal{D}(\mathcal{T})$. A dispute tree $\mathcal{T}$ is an *admissible dispute tree* iff:

1. every O node in $\mathcal{T}$ has a child, and
2. no argument in $\mathcal{T}$ labels both a P and an O node.

We have the result:

1. If $\mathcal{T}$ is an admissible dispute tree for an argument A, then $\mathcal{D}(\mathcal{T})$ is admissible.
2. If A is an argument and $A \in E$ where E is an admissible set then there exists an admissible dispute tree for A with $\mathcal{D}(\mathcal{T}) = E'$ such that $E' \subseteq E$ and E' is admissible.

---

^{5} $\tau \notin \mathcal{L}$ represents "true" and stands for the empty body of rules.

5.2 Decision Graphs

Multi-criteria decision making problems can be modelled with *decision graphs* where

- nodes may be *decisions*, *goals*, or *intermediates*;
- edges represent relations amongst nodes, e.g. an edge from a decision to an intermediate represents that the decision *has* the attribute; an edge from an intermediate to a goal represents that the intermediate *satisfies* the goal; an edge from an intermediate to another intermediate represents that the former *leads to* the latter;
- edges are equipped with *tags*.

We assume that all decision graphs are acyclic, and that tags are natural numbers. Different edges may have the same tag.

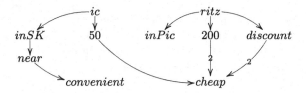

Fig. 8. An example decision graph

As a simple example, consider an agent who needs to decide on accommodation in London. The agent wants this accommodation to be convenient and cheap. The two candidates are Imperial College London Student Accommodation (*ic*) and Ritz Hotel (*ritz*). *ic* is £50 a night and in South Kensington (*inSK*). Ritz is normally £200 a night and in Piccadilly (*inPic*). However, Ritz is having a promotion discount. Hence both accommodations are cheap. However, South Kensington is near so it is convenient whereas Piccadilly is not. Intuitively, *ic* is the better choice between the two. This decision problem can be represented as the decision graph in Fig. 8. Here

- the decisions are: $D = \{ic, ritz\}$;
- the goals are: $G = \{convenient, cheap\}$;
- the intermediates are: $I = \{inSK, 50, inPic, 200, discount, near\}$;
- all edges are (implicitly) tagged 1, except for the edges explicitly tagged 2.

We can define a notion of *reachability* from a set of nodes to a node as follows:

C1. if there is a set of nodes N all having edges with the same tag pointing to some node n, then n is reachable from N;
C2. if there is some set of nodes N' satisfying $C1$ (n being reachable from N') and every single node n' in N' is reachable from N by satisfying either $C1$ or $C2$, then n is reachable from N.

This is a recursive definition, with C1 the base case.

For the decision graph in Fig. 8:

- $inSK$ and 50 are reachable from $\{ic\}$,
- $inPic$, 200, $discount$ are reachable from $\{ritz\}$,
- $near$ is reachable from $\{inSK\}$ and $\{ic\}$,
- $convenient$ is reachable from $\{near\}$, $\{inSK\}$, $\{ic\}$,
- $cheap$ is reachable from $\{50\}$, $\{200, discount\}$, $\{ic\}$ and $\{ritz\}$.

Then, a notion of decisions *meeting* goals can be given in terms of reachability of the goal from the singleton set consisting of the decision. Thus, for the example in Fig. 8, the decision ic meets the goals *convenient* and *cheap*; and the decision $ritz$ only meets the goal *cheap*.

Finally, following [26], a decision d can be deemed to be *dominant* if there is no goal g such that d does not meet g but there exists some d' meeting g. In short, a decision is dominant iff it meets all goals that can be met. For the decision graph in Fig. 8, since ic meets *convenient* and *cheap* and $ritz$ meets *cheap*, ic meets all goals that are ever met. Therefore ic is dominant.

Dominance can then be used as the decision criterion by means of which to choose decisions. (Other decision criteria can be defined, e.g. as in [20,21]; we focus here on dominance by way of illustration).

5.3   Dominant Decisions in Decision Graphs in ABA

The problem of identifying dominant decisions in decision graphs can be understood equivalently as the problem of identifying admissible arguments in ABA. This in turn allows a further graphical representation, incorporating the decision problem and the decision criterion at the same time, that can be used to explain, via debate trees, best decisions.

As an example, given the decision framework in Fig. 8, the ABA framework is $\langle \mathcal{L}, \mathcal{R}, \mathcal{A}, \mathcal{C} \rangle$ where:

- $\mathcal{R}$ consists of:

  $isD(ic) \leftarrow$ $isD(ritz) \leftarrow$

  $isG(convenient) \leftarrow$ $isG(cheap) \leftarrow$

  $edge(ic, inSK, 1) \leftarrow$ $edge(ic, 50, 1) \leftarrow$

  $edge(ritz, inPic, 1) \leftarrow$ $edge(ritz, 200, 1) \leftarrow$

  $edge(ritz, discount, 1) \leftarrow$ $edge(inSK, near, 1) \leftarrow$

  $edge(near, convenient, 1) \leftarrow$ $edge(50, cheap, 1) \leftarrow$

  $edge(discount, cheap, 2) \leftarrow$ $edge(200, cheap, 2) \leftarrow$

as well as all instances of the following rule schemata, with variables X, Y, Z, W instantiated to elements of $D \cup G \cup I$, variable S instantiated to elements of G and variable T instantiated to tags in the decision graph:

$reach(X, Y) \leftarrow edge(X, Y, T)$

$reach(X, Y) \leftarrow reach(X, Z), edge(Z, Y, T), \neg unreachableSib(Z, Y, T, X)$

$unreachableSib(Z, Y, T, X) \leftarrow edge(W, Y, T), W {\neq} Z, \neg reach(X, W)$

$met(D, S) \leftarrow reach(D, S), isD(D), isG(S)$

$notDom(D) \leftarrow notMet(D, S), isD(D), isG(S)$

$othersMet(D, S) \leftarrow met(D_1, S), D \neq D_1$

- $\mathcal{A} = \mathcal{A}^{\neg u} \cup \mathcal{A}^{\neg r} \cup \mathcal{A}^{rest}$ where, for $\mathcal{S} = \{ic, ritz, inSK, 50, inPic, 200,$
  $discount, near, convenient, cheap\}$:
  $\mathcal{A}^{\neg u} = \{\neg unreachableSib(n', n'', t, n)|$

      $n \in \mathcal{S}$,
  $(n', n'') \in \{[ic, inSK], [ic, 50], [ritz, inPic], [ritz, 200], [ritz, discount],$
      $[inSK, near], [50, cheap], [200, cheap], [discount, cheap],$
      $[near, convenient]\}, n \neq n', n \neq n'',$
      and t is the tag of $(n', n'')\}$
  $\mathcal{A}^{\neg r} = \{\neg reach(n, n')|\ n, n' \in \mathcal{S},\ n \neq n'\}$
  and $\mathcal{A}^{rest}$ consists of:

$dom(ic)$	$dom(ritz)$
$notMat(ic, convenient)$	$notMat(ic, cheap)$
$notMat(ritz, convenient)$	$notMat(ritz, cheap)$
$noOthers(ic, convenient)$	$noOthers(ic, cheap)$
$noOthers(ritz, convenient)$	$noOthers(ritz, cheap)$

- $\mathcal{C}$ is such that (for all instances of the variables as for $\mathcal{R}$):
  $\mathcal{C}(\neg unreachableSib(Z, Y, T, X)) = \{unreachableSib(Z, Y, T, X)\}$;
  $\mathcal{C}(\neg reach(X, Y)) = \{reach(X, Y)\}$;
  $\mathcal{C}(dom(D)) = \{notDom(D)\}$;
  $\mathcal{C}(notMet(D, S)) = \{met(D, S), noOthers(D, S)\}$;
  $\mathcal{C}(noOthers(D, S)) = \{othersMet(D, S)\}$.

Dominant decisions in the decision graph G correspond then to admissible arguments in the ABA framework. Formally: d is dominant in G iff $\{dom(d)\} \vdash dom(d)$ is admissible in the ABA framework.

In our running example, argument $\{dom(ic)\} \vdash dom(ic)$ is admissible and ic is dominant. The admissible dispute tree for this argument, shown in Fig. 9, explains why ic is indeed dominant: the root shows the claim that ic is a dominant decision; this claim is challenged by two opponent arguments, A and B, stating that ic does not meet the goals $convenient$ and $cheap$, respectively; A and B are counter-attacked by C and D, respectively, confirming that ic meets both goals.

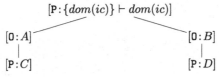

$A = \{notMet(ic, convenient)\} \vdash notDom(ic)$
$B = \{notMet(ic, cheap)\} \vdash notDom(ic)$
$C = \{\neg unreachableSib(near, convenient, 1, ic), \neg unreachableSib(inSK, near, 1, ic)\} \vdash$
$met(ic, convenient)$
$D = \{\neg unreachableSib(50, cheap, 1, ic)\} \vdash met(ic, cheap)$.

Fig. 9. The dispute tree for $\{dom(ic)\} \vdash dom(ic)$.

6   Discussion and Future Directions

We have given an overview of various uses of Argumentation Frameworks of various kinds to support some forms of decision-making. In addition to affording a graphical view of decision problems, argumentation provides the capability to evaluate the graphs, with the evaluation being linked to best decisions, in the different ways considered. To conclude, we discuss various directions for future research in this area.

6.1   From Trees to Full Graphs

The current structure of debates both in QuAD and ESAA (see Sect. 3) currently only allows the evaluation of tree structures. It will be important to broaden the class of debates that can be analysed, extending the current evaluation functionality to cope with the evaluation of directed graphs of more general forms. In terms of allowing more generic graph structures and visualising them in Quaestio-it, for example, the main issue lies not with the debate process, but rather with the evaluation of argument strength values.

6.2   Constructing Bipolar Argumentation Frameworks from Text

Much work in Knowledge Representation has traditionally struggled facing a knowledge acquisition bottleneck. In this, argumentation research is no exception; much work needs to be invested in constructing graphs of arguments that can then be used as discussed earlier. One way to remedy this issues may be presented by Argumentation Mining (AM) [27], a fairly young discipline in Natural Language Processing, aiming to identify arguments, and their relations to each other, in natural language text. One possible AM approach [7] is to construct BAFs from text, where the nodes in the BAF graphs constitute *argumentative* sentences and edges between the nodes signify attack and support relations between sentences. To achieve this by means of supervised learning techniques, we need to construct corpora of the form

$$P = \{(s_1, s_1', C_1), (s_2, s_2', C_2), \ldots, (s_n, s_n', C_3)\},$$

where each s_i, s_i' is a chunk of text and C_i is a class label in $\{A, S, N\}$, for $A = $ Attack, $S = $ Support and $N = $ Neither. Basically, each triple in P determines whether sentence s_i attacks, supports or neither attacks nor supports s_i'. Rather than determining whether a sentence is an argument in its own right, this approach aims to determine whether it has argumentative qualities in relation to another sentence. For new pairs of sentences (s, s'), a trained classifier can then be used to determine the relation between s and s', and, if this is attack or support, help construct a graph of arguments, as in bipolar abstract argumentation. Apart from building a reliable classification model to label sentence pairs the main challenge faced here is that of finding a way of constructing sentence pairs that is not computationally prohibitive. The approach described in

[7] foregoes this issue by selecting a fixed *Hypothesis* against which to evaluate the contents of a text. We may, for example, propose a hypothesis on whether a certain politician is fit to be the leader of his or her party:

> s_a = Mrs. Johnson is hardly qualified to lead *the Party*, being that she is both inexperienced and has had some dubious dealings with prominent business figures.

We may then construct sentence pairs by matching our hypothesis with sentences from news articles that discuss said politician, e.g.

> Looking at her voting record on recent bills she has also been accused of pandering to certain industries which she has allegedly invested in.

Doing so for every sentence in a collection of relevant sentences gives us an idea of how the text agrees or disagrees with our hypothesis; it does not, however, enable us to construct the desired graph. To achieve this, we need to construct sentence pairs from the text itself. When analysing text that is comprised of more than just a handful of sentences, building sentence pairs by combining every sentence with every other sentence in the text is generally not an option, as the amount of sentence pairs we need to analyse grows quadratically. We hence need to find ways of preselecting a set of candidate sentences from which we construct the pairs.

6.3   Computation of Acceptability/Strength

The computation of strength in Bipolar Argumentation is simple enough for trees, as Quaestio-it shows. However, once graphs are accommodated for, the computational burden will increase. Much work has been devoted to computational issues in argumentation. For example, [10,15–17] give complexity results for several AF types and reasoning problems. It is generally recognised that computation in argumentation is expensive. Novel algorithms have been proposed to address computational challenges in argumentation: [31] represent AFs in propositional satisfiability (SAT) programs and rely on advanced SAT solvers for computing acceptable arguments; [2] propose to understand various argumentation semantics as Strongly Connected Component concepts, yielding new semantics computation methods such as the one in [9]; [25] give an overview of recent development in argumentation semantics computation; several computational mechanisms for ABA have been proposed (e.g. see [13,30]). The impact of different computational mechanisms on decision-making applications is an important issue to be studied further.

6.4   Cross-Fertilisations

DesMOLD uses relevant past debates to automatically extend an ongoing debate, as described in Sect. 3.3. It would be useful to integrate arguments from ABA frameworks, e.g. to model policy rules in a specific design domain within debates. Collaborative Multi-Agent decision making (see Sect. 4) uses qualitative evaluation of arguments (using for example admissibility) in the presence solely of an attack relation. The use of supporting arguments, as in Bipolar Argumentation, and of quantitative evaluation methods, may also be useful in this setting.

Acknowledgements. This research was supported by the EPSRC TRaDAr project *Transparent Rational Decisions by Argumentation*: EP/J020915/1 and the EU project *DesMOLD (FP7/2007-2013-314581)*.

References

1. Amgoud, L.: Argumentation for decision making. In: Rahwan, I., Simari, G.R. (eds.) Argumentation in AI, pp. 301–320. Springer, Heidelberg (2009)
2. Baroni, P., Giacomin, M., Guida, G.: SCC-recursiveness: a general schema for argumentation semantics. Artif. Intell. **168**(1–2), 162–210 (2005)
3. Baroni, P., Romano, M., Toni, F., Aurisicchio, M., Bertanza, G.: An argumentation-based approach for automatic evaluation of design debates. In: Leite, J., Son, T.C., Torroni, P., van der Torre, L., Woltran, S. (eds.) CLIMA XIV 2013. LNCS, vol. 8143, pp. 340–356. Springer, Heidelberg (2013)
4. Baroni, P., Romano, M., Toni, F., Aurisicchio, M., Bertanza, G.: Automatic evaluation of design alternatives with quantitative argumentation. Argument Comput. **6**(1), 24–49 (2015)
5. Bench-Capon, T.J.M.: Persuasion in practical argument using value-based argumentation frameworks. J. Logic Comput. **13**(3), 429–448 (2003)
6. Cabanillas, D., Bonada, F., Ventura, R., Toni, F., Evripidou, V., Carstens, L., Rebolledo, L.: A combination of knowledge and argumentation based system for supporting injection mould design. In: Gibert, K., Botti, V.J., Bolaño, R.R. (eds.) Proceedings of ICCAAI. Frontiers in Artificial Intelligence and Applications, vol. 256, pp. 293–296. IOS Press, Amsterdam (2013)
7. Carstens, L., Toni, F.: Towards relation based argumentation mining. In: The 2nd Workshop on Argumentation Mining, NAACL (2015)
8. Cayrol, C., Lagasquie-Schiex, M.C.: On the acceptability of arguments in bipolar argumentation frameworks. In: Godo, L. (ed.) ECSQARU 2005. LNCS (LNAI), vol. 3571, pp. 378–389. Springer, Heidelberg (2005)
9. Cerutti, F., Giacomin, M., Vallati, M., Zanella, M.: An SCC recursive meta-algorithm for computing preferred labellings in abstract argumentation. In: Proceedings of KR (2014)
10. Dimopoulos, Y., Nebel, B., Toni, F.: On the computational complexity of assumption-based argumentation for default reasoning. Artif. Intell. **141**(1/2), 57–78 (2002)
11. Dung, P.M.: On the acceptability of arguments and its fundamental role in non-monotonic reasoning, logic programming and n-person games. Artif. Intell. **77**(2), 321–357 (1995)
12. Dung, P.M., Kowalski, R.A., Toni, F.: Dialectic proof procedures for assumption-based, admissible argumentation. Artif. Intell. **170**(2), 114–159 (2006)
13. Dung, P.M., Kowalski, R.A., Toni, F.: Assumption-based argumentation. In: Rahwan, I., Simari, G.R. (eds.) Argumentation in AI, pp. 199–218. Springer, Heidelberg (2009)

14. Dung, P.M., Mancarella, P., Toni, F.: Computing ideal sceptical argumentation. Artif. Intell. **171**(10–15), 642–674 (2007)
15. Dunne, P.E.: The computational complexity of ideal semantics. Artif. Intell. **173**(18), 1559–1591 (2009)
16. Dunne, P.E., Caminada, M.: Computational complexity of semi-stable semantics in abstract argumentation frameworks. In: Hölldobler, S., Lutz, C., Wansing, H. (eds.) JELIA 2008. LNCS (LNAI), vol. 5293, pp. 153–165. Springer, Heidelberg (2008)
17. Dunne, P.E., Hunter, A., McBurney, P., Parsons, S., Wooldridge, M.: Weighted argument systems: basic definitions, algorithms, and complexity results. Artif. Intell. **175**(2), 457–486 (2011)
18. Evripidou, V., Carstens, L., Toni, F., Cabanillas, D.: Argumentation-based collaborative decisions for design. In: ICTAI 2014, pp. 805–809. IEEE (2014)
19. Evripidou, V., Toni, F.: Quaestio-it.com - a social intelligent debating platform. J. Decis. Syst. **23**(3), 333–349 (2014)
20. Fan, X., Craven, R., Singer, R., Toni, F., Williams, M.: Assumption-based argumentation for decision-making with preferences: a medical case study. In: Leite, J., Son, T.C., Torroni, P., van der Torre, L., Woltran, S. (eds.) CLIMA XIV 2013. LNCS, vol. 8143, pp. 374–390. Springer, Heidelberg (2013)
21. Fan, X., Toni, F.: Decision making with assumption-based argumentation. In: Modgil, S., Oren, N., Black, E. (eds.) TAFA 2013. LNCS, vol. 8306, pp. 127–142. Springer, Heidelberg (2014)
22. Gao, Y., Toni, F.: Argumentation accelerated reinforcement learning for cooperative multi-agent systems. In: Proceedings of ECAI (2014)
23. Kapetannakis, S., Kudenko, D.: Reinforcement learning of coordination in cooperative multi-agent systems. In: Proceedings of AAAI (2002)
24. Leite, J., Martins, J.: Social abstract argumentation. In Proceedings of IJCAI (2011)
25. Liao, B.: Efficient Computation of Argumentation Semantics. Academic Press, Oxford (2014)
26. Matt, P.-A., Toni, F., Vaccari, J.R.: Dominant decisions by argumentation agents. In: McBurney, P., Rahwan, I., Parsons, S., Maudet, N. (eds.) ArgMAS 2009. LNCS, vol. 6057, pp. 42–59. Springer, Heidelberg (2010)
27. Palau, R.M., Moens, M.: Argumentation mining: the detection, classification and structure of arguments in text. In: Proceedings of ICAIL, pp. 98–107 (2009)
28. Rahwan, I., Simari, G.R. (eds.): Argumentation in AI. Springer, Heidelberg (2009)
29. Russell, S., Norvig, P.: Artificial Intelligence: A Modern Approach. Prentice Hall, Upper Saddle River (2009)
30. Toni, F.: A tutorial on assumption-based argumentation. Argument Comput. **5**(1), 89–117 (2014)
31. Wallner, J.P., Weissenbacher, G., Woltran, S.: Advanced SAT techniques for abstract argumentation. In: Leite, J., Son, T.C., Torroni, P., van der Torre, L., Woltran, S. (eds.) CLIMA XIV 2013. LNCS, vol. 8143, pp. 138–154. Springer, Heidelberg (2013)

Learning Bayesian Networks
with Non-Decomposable Scores

Eunice Yuh-Jie Chen[(✉)], Arthur Choi, and Adnan Darwiche

Computer Science Department, University of California, Los Angeles, USA
{eyjchen,aychoi,darwiche}@cs.ucla.edu

Abstract. Modern approaches for optimally learning Bayesian network structures require decomposable scores. Such approaches include those based on dynamic programming and heuristic search methods. These approaches operate in a search space called the order graph, which has been investigated extensively in recent years. In this paper, we break from this tradition, and show that one can effectively learn structures using non-decomposable scores by exploring a more complex search space that leverages state-of-the-art learning systems based on order graphs. We show how the new search space can be used to learn with priors that are not structure modular (a particular class of non-decomposable scores). We also show that it can be used to efficiently enumerate the k-best structures, in time that can be up to three orders of magnitude faster, compared to existing approaches.

1 Introduction

Modern approaches for learning Bayesian network structures are typically formulated as a (combinatorial) optimization problem, where one wants to find the best network structure (i.e., best DAG) that has the highest score, for some given scoring metric [10,19,23]. Typically, one seeks a Bayesian network that explains the data well, without overfitting the data, and ideally, also accommodating any prior knowledge that may be available.

Some of the earliest procedures for learning Bayesian network structures used scoring metrics with a certain desirable property, called *score decomposability*. For example, consider the K2 algorithm which exploited decomposable scores, in combination with an assumption on the topological ordering of the variables [6]. Under these assumptions, the structure learning problem itself decomposes into local sub-problems, where we find the optimal set of parents for each variable, independently. Local search methods exploited decomposability in a similar way [5]. Such methods navigated the space of Bayesian network structures, using operators on edges such as addition, deletion, and reversal. Score decomposability allowed these operators to be evaluated locally and efficiently. Indeed, almost all scoring metrics used for Bayesian network structure learning are decomposable. Such scores include the K2 score, [6], the BDeu score [4], the BDe score [16], and the MDL score [2,20,29], among many others.

© Springer International Publishing Switzerland 2015
M. Croitoru et al. (Eds.): GKR 2015, LNAI 9501, pp. 50–71, 2015.
DOI: 10.1007/978-3-319-28702-7_4

Modern approaches to structure learning continue to exploit the decomposable nature of such scoring metrics. In particular, the past decade has seen significant developments in *optimal* Bayesian network structure learning. These recent advances were due in large part to dynamic programming (DP) algorithms, for finding optimal Bayesian network structures [18,27,28]. Subsequently, approaches have been proposed based on heuristic search, such as A* search [34,35], as well as approaches based on integer linear programming (ILP), and their relaxations [7,17].

By exploiting the nature of decomposable scores, these advances have significantly increased the scalability of optimal Bayesian network structure learning. There is, however, a notable void in the structure learning landscape due to the relative lack of support for *non-decomposable scores*. This includes a general lack of support for non-decomposable priors, or more broadly, the ability to incorporate more expressive, but non-decomposable forms of prior knowledge (e.g., biases or constraints on ancestral relations). In this paper, we take a step towards a more general framework for Bayesian network structure learning that targets this void.

The modern approaches for optimal structure learning, mentioned earlier, are based on a search space called the order graph [18,35]. The key property of the order graph is its size, which is only exponential in the number of variables of the Bayesian network that we want to learn. Our proposed framework however is based on navigating the significantly larger space of all network structures (i.e., all DAGs). Moreover, to facilitate the efficient navigation of this larger space, we employ state-of-the-art learning systems based on order graphs as a (nearly) omniscient oracle. In addition to defining this new search space, we instantiate it to yield a concrete system for finding optimal Bayesian networks under order-modular priors, which we evaluate empirically. We further demonstrate the utility of this new search space by showing how it lends itself to enumerating the k-best structures, resulting an algorithm that can be three orders of magnitude more efficient than existing approaches based on DP and ILP [9,32].

This paper is organized, as follows. In Sect. 2, we review Bayesian network structure learning. In Sect. 3, we propose our new search space for learning Bayesian networks. In Sect. 4, we show how our search space can be leveraged to find optimal Bayesian networks under a class of non-decomposable priors. In Sect. 5, we show how our search space can be further used to efficiently enumerate the k-best network structures. Finally, we conclude in Sect. 6. Proofs of theorems are provided in the Appendix.

2 Technical Preliminaries and Related Work

In this section, we first review score-based Bayesian network structure learning. We then review a formulation of score-based structure learning as a shortest-path problem in a graph called the order graph. Shortest-path problems can subsequently be solved with heuristic search methods such as A* search.

First, we use upper case letters (X) to denote variables and bold-face upper case letters $(\mathbf{X})$ to denote sets of variables. Generally, we will use X to denote a variable in a Bayesian network and $\mathbf{U}$ to denote its parents.

2.1 Score-Based Structure Learning

Given a dataset $\mathcal{D}$, we first consider the problem of finding a DAG G of a Bayesian network which minimizes a *decomposable score*. Such a score decomposes into a sum of local scores, over the families $X\mathbf{U}$ of the DAG:

$$\mathsf{score}(G \mid \mathcal{D}) = \sum_{X\mathbf{U}} \mathsf{score}(X\mathbf{U} \mid \mathcal{D}). \tag{1}$$

For example, MDL and BDeu scores are decomposable; see, e.g., [10,19,23]. In this paper, we will generally assume that scores (costs) are to be minimized (hence, we negate scores that should otherwise be maximized).

There are a variety of approaches for finding a DAG G that minimizes the score of Eq. 1. One class of approaches is based on integer linear programming (ILP), where 0/1 variables represent the selection of parent sets (families) in a graph. Our goal is to optimize the (linear) objective function of Eq. 1, subject to (linear) constraints that ensure that the resulting graph is acyclic [17]. In some cases, an LP relaxation can guarantee an optimal solution to the original ILP; otherwise, cutting planes and branch-and-bound algorithms can be used to obtain an optimal structure [7,17].

In this paper, we are interested in another class of approaches to optimizing Eq. 1, which is based on a formulating the score, in a particular way, as a recurrence. This recurrence underlies a number of recent approaches to structure learning, based on dynamic programming [18,22,27,28], as well as more efficient approaches based on A* search [34,35]. In particular, to find the optimal DAG over variables $\mathbf{X}$, we have the following recurrence:

$$\mathsf{score}^\star(\mathbf{X} \mid \mathcal{D}) = \min_{X \in \mathbf{X}} \left(\min_{\mathbf{U} \subseteq \mathbf{X}\setminus X} \mathsf{score}(X\mathbf{U} \mid \mathcal{D}) + \mathsf{score}^\star(\mathbf{X} \setminus X \mid \mathcal{D}) \right) \tag{2}$$

where $\mathsf{score}^\star(\mathbf{X} \mid \mathcal{D})$ denotes the score of the optimal DAG over variables $\mathbf{X}$ given dataset $\mathcal{D}$. According to this recurrence, we evaluate each variable X as a candidate leaf node, and find its optimal family $X\mathbf{U}$. Moreover, independently, we find the optimal structure over the remaining variables $\mathbf{X} \setminus X$. The best structure then corresponds to the candidate leaf node X with the best score.

2.2 Shortest-Paths on Order Graphs

Yuan & Malone [34] formulate the structure learning problem as a shortest-path problem on a graph called the *order graph*. Figure 1 illustrates an order graph over 3 variables $\mathbf{X}$. In an order graph, each node represents a subset $\mathbf{Y}$ of the variables $\mathbf{X}$. There is a directed edge from $\mathbf{Y}$ to $\mathbf{Z}$ in the order graph iff we add a new variable X to the set $\mathbf{Y}$, to obtain the set $\mathbf{Z}$; we denote such an edge by

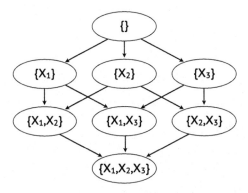

Fig. 1. An order graph for variables $\mathbf{X} = \{X_1, X_2, X_3\}$

$\mathbf{Y} \xrightarrow{X} \mathbf{Z}$. The order graph is thus a layered graph where each new layer adds, in total, one new variable to the nodes of the previous layer. Hence, we have a unique root node $\{\}$, and a unique leaf node $\mathbf{X}$. Any path

$$\{\} \xrightarrow{X_1} \cdots \xrightarrow{X_n} \mathbf{X}$$

from the root to the leaf will then correspond to a unique ordering $\langle X_1, \ldots, X_n \rangle$ of the variables. Suppose that we associate each edge $\mathbf{Y} \xrightarrow{X} \mathbf{Z}$ with a cost

$$\min_{\mathbf{U} \subseteq \mathbf{Y}} \mathsf{score}(X\mathbf{U} \mid \mathcal{D})$$

where, for the variable X added on the edge, we find the optimal set of parents $\mathbf{U}$ from the set of variables $\mathbf{Y}$. A path from the root node $\{\}$ to the leaf node $\mathbf{X}$ will then correspond to a DAG G since each edge $\mathbf{Y} \xrightarrow{X} \mathbf{Z}$ adds a new leaf node X with parents $\mathbf{U}$ to the DAG, i.e., the $\mathbf{U}$ that minimized $\mathsf{score}(X\mathbf{U} \mid \mathcal{D})$. The cost of the path (the sum of the scores of its edges) gives us the score of the DAG, $\mathsf{score}(G \mid \mathcal{D})$, as in Eqs. 1 and 2. Hence, the shortest path from the root $\{\}$ to the leaf $\mathbf{X}$ corresponds to the DAG with minimum score.

3   A New Search Space for Learning Bayesian Networks

We now describe our A* framework, for learning the structure of Bayesian networks. We first describe the search space that we use, and then propose a heuristic function to navigate that space, which is based on using existing, state-of-the-art structure learning systems as a black-box. Later in this paper, we discuss two learning tasks that are enabled by our framework: (1) learning an optimal Bayesian network structure using a class of non-decomposable scores, and (2) enumerating the k-best Bayesian network structures.

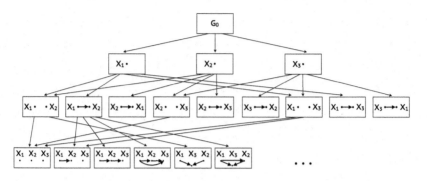

Fig. 2. A BN graph for variables $\mathbf{X} = \{X_1, X_2, X_3\}$

3.1 A New Search Space: BN Graphs

Following Yuan & Malone [34], we formulate structure learning as a shortest-path problem, but on a different graph, which we call the *Bayesian network graph (BN graph)*. The BN graph is a graph where each node represents a BN, but more specifically, each node represents a BN structure, i.e., a DAG. Figure 2 illustrates a BN graph over 3 variables. In this graph, which we denote by $\mathcal{G}_{\text{bn}}$, nodes represent Bayesian network structures over different subsets of the variables $\mathbf{X}$. A directed edge $G_i \xrightarrow{X\mathbf{U}} G_j$ from a DAG G_i to a DAG G_j exists in $\mathcal{G}_{\text{bn}}$ iff G_j can be obtained from G_i by introducing variable X as a leaf node with parents $\mathbf{U}$. Hence, the BN graph, like the order graph, is a layered graph, but where each layer adds one more leaf to an explicit (and not just an implicit) DAG when we walk an edge to the next layer. Correspondingly, when we refer to a DAG G_i, we assume it is on the i-th layer, i.e., G_i has i nodes. The top (0-th) layer contains the root of the BN graph, a DAG with no nodes, which we denote by G_0. The bottom (n-th) layer contains DAGs G_n over our n variables $\mathbf{X}$. Any path

$$G_0 \xrightarrow{X_1\mathbf{U}_1} \cdots \xrightarrow{X_n\mathbf{U}_n} G_n$$

from the root to a DAG G_n on the bottom layer, is a construction of the DAG G_n, where each edge $G_{i-1} \xrightarrow{X_i\mathbf{U}_i} G_i$ adds a new leaf X_i with parents $\mathbf{U}_i$. Moreover, each path corresponds to a unique ordering $\langle X_1, \ldots, X_n \rangle$ of the variables. Each edge $G_{i-1} \xrightarrow{X_i\mathbf{U}_i} G_i$ is associated with a cost, $\text{score}(X_i\mathbf{U}_i \mid \mathcal{D})$, and thus the cost of a path from the empty DAG G_0 to a DAG G_n gives us the score of the DAG, $\text{score}(G_n \mid \mathcal{D})$, as in Eq. 1.

For example, consider the BN graph of Fig. 2 and the following path, corresponding to a sequence of DAGs:

G_0	G_1	G_2	G_3
	X_1	$X_1 \to X_2$	$X_1 \to X_2 \quad X_3$

Starting with the empty DAG G_0, we add a leaf X_1 (with no parents), then a leaf X_2 (with parent X_1), then a leaf X_3 (with no parents), giving us a DAG G_3 over all 3 variables.

Both the order graph and the BN graph formulate the structure learning problem as a shortest path problem. The BN graph is however much larger than the order graph: an order graph has 2^n nodes, whereas the BN graph has $O(n! \cdot 2^{\binom{n}{2}})$ nodes. Despite this significant difference in search space size, we are still able to efficiently find shortest-paths in the BN graph, which we shall illustrate empirically, in the remainder of this paper. The efficient navigation of the BN graph depends significantly on the heuristic function, which we discuss next.

3.2   On Heuristic Functions for BN Graphs

A* search is a best-first search that uses an *evaluation* function f to guide the search process, where we expand first those nodes with the lowest f cost [15]. The evaluation function for A* takes the form:

$$f(G) = g(G) + h(G)$$

where G is a given DAG, function g is the *path cost* (the cost of the path to reach G from G_0), and function h is the *heuristic function*, which estimates the cost to reach a goal, starting from G. If our heuristic function h is *admissible*, i.e., it never over-estimates the cost to reach a goal, then A* search is optimal. That is, the first goal node G_n that A* expands is the one that has the shortest path from the root G_0. Ideally, we want a good heuristic function h, since an accurate estimate of the cost to a goal state will lead to a more efficient search. For more on A* search, see, e.g., [26].

Consider the special but extreme case, where we have access to a *perfect* heuristic $h(G)$, which could predict the optimal path from G to a goal node G_n. In this case, search becomes trivial: A* search marches straight to a goal node (with appropriate tie-breaking, where we expand the deepest node first). Having access to a perfect heuristic by itself is not useful, if we are just interested in an optimal DAG. Such a heuristic, however, becomes useful when we are interested in solving more challenging learning tasks. Consider, for example, learning an optimal DAG, subject to a set of structural constraints. In this case, a perfect heuristic is no longer perfect—it will under-estimate the cost to reach a goal. However, in this case, it remains an admissible heuristic, which we can use in A* search to find an optimal DAG, when we subject the learning problem to constraints.

We do, in fact, have access to a perfect heuristic—any learning system could be used as such, provided that it can accept a (partial) DAG G, and find an optimal DAG G_n that extends it. Systems such as URLEARNING meet this criterion [34], which we use in our subsequent experiments. Such a system is treated as a black-box, and used to evaluate our heuristic function in A* search, to potentially solve a learning problem that the black-box was not originally designed for. We shall later highlight two such learning tasks, that are enabled by using existing structure learning systems as black-boxes for A* search.

We also remark that using such a black-box to evaluate a heuristic function, as described above, is also a departure from the standard practice of heuristic search. Conventionally, in heuristic search, one seeks heuristic functions that are cheap to evaluate, allowing more nodes to be evaluated, and hence more of the search space to be explored. Our black-box (which finds an optimal extension of a DAG), in contrast, will be relatively expensive to evaluate. However, for the particular learning tasks that we consider, a strong heuristic can outweigh the expense to compute it, by more efficiently navigating the search space (i.e., by expanding fewer nodes to find a goal). We shall demonstrate this empirically after introducing each of the learning tasks that we consider.

Implementation of A* Search. Finally, we describe two further design decisions, that are critical to the efficiency of A* search on the BN graph. First, if two given DAGs G and G' are defined over the same set of variables, then they have the same heuristic value, i.e. $h(G) = h(G')$. Hence, we can cache the heuristic value $h(G)$ for a DAG G, and simply fetch this value for another DAG G' (instead of re-invoking our black-box), when it has the same set of variables. As a result, the heuristic function is invoked at most once for each subset **Y** of the variables **X**. In addition, when we invoke our black-box on a DAG G, we can infer and then prime other entries of the cache. In particular, when our black-box returns an optimal completion G' of a DAG G, then we know the optimal completion (and heuristic values) of any DAG in between G and G' in the BN graph—their optimal completion is also G' (from which we can infer the corresponding heuristic value). Based on this caching scheme, a single call to our black-box heuristic function suffices, to recover a single best network using A* search in the BN graph (i.e., it is no worse than using the black-box directly).

Next, the branching factor of the BN graph is large, and hence, we can quickly run out of memory if we expand each node and insert all of its children into A*'s priority queue (i.e., open list). We thus use partial-expansion A* in our implementation, i.e., when we expand a node, we only insert the b-best children into the priority queue. We can re-expand this node, as many times as needed, when we want the next b-best children. While we may spend extra work re-expanding nodes, this form of partial-expansion can save a significant amount of memory, without sacrificing the optimality of A* search; see, e.g., [13,33].

3.3   Experimental Setup

In the subsequent sections, we shall highlight two different tasks that are enabled by performing A* search on the BN graph. After discussing each task, we report empirical results on real-world datasets, which were taken from the UCI machine learning repository [1], and the National Long Term Care Survey (NLTCS). For learning, we assumed BDeu scores, with an equivalent sample size of 1. We adapted the URLEARNING structure learning package of [34] to serve as our black-box heuristic function.[1] Our experiments were run on a 2.67GHz Intel

---

[1] At https://sites.google.com/site/bmmalone/files/urlearning.

Xeon X5650 CPU, with access to 144 GB RAM. For our partial-expansion A* search, each time a node is expanded or re-expanded, children are inserted into the priority queue in blocks of 10. We further pre-compute the BDeu scores, which are fed as input into each system evaluated. Finally, all timing results are averages over 10 runs.

4 Structure Learning with Non-Decomposable Scores

Now that we have described our framework for learning Bayesian networks using BN graphs, we will show how we can use it to learn BN structures using for a class of non-decomposable scores.[2] In particular, we consider a class of non-decomposable priors on network structures, which we discuss next. Subsequently, we propose a heuristic that can be used in A* search, to optimize this class of non-decomposable scores. We then describe our A* search algorithm, and then provide some experimental results.

4.1 Order-Modular Priors

One prevalent non-decomposable prior is the *order-modular* prior [14,18]. The *uniform* order-modular prior $Pr(G)$, in particular, is proportional to the number of topological orderings consistent with a DAG G, i.e., the number of its linear extensions, which we denote by $\#G$. Hence,

$$\log Pr(G) = \log \#G - \log C,$$

where C is a normalizing constant. More generally, order-modular priors can be viewed in terms of a *weighted* count of linear extensions [18]. In general, counting the number of linear extensions is itself a challenging problem (let alone optimizing with it); it is a #P-complete problem [3]. We shall revisit this issue, shortly.

Order-modular priors are notable, as they enable MCMC methods for the purposes of (approximate) Bayesian model averaging [14]. They also enable some DP-based methods for exact Bayesian model averaging, when there are a moderate number of network variables [18]. However, to our knowledge, only approximate approaches had been previously considered for this prior, when one wants a single optimal DAG; see Koivisto & Sood [18, Sect. 5], for a discussion on some of the difficulties.

4.2 A Heuristic for Order-Modular Priors

Consider the probability of a DAG G given a dataset $\mathcal{D}$:

$$Pr(G \mid \mathcal{D}) = \frac{Pr(\mathcal{D} \mid G)Pr(G)}{Pr(\mathcal{D})}$$

---

[2] Approaches based on ILP can in principle handle non-decomposable scores (and constraints), assuming that they can be expressed using a linear cost function (or as linear constraints) [25]. We remark that order-modular priors, which we consider later, are not easy to encode as ILPs (as we need to compute linear extension counts).

where $Pr(\mathcal{D} \mid G)$ is the marginal likelihood, and $Pr(G)$ is a prior on the DAG G. Further, the quantity $Pr(\mathcal{D})$ is a normalizing constant, which is independent of the given DAG G. To maximize the probability of a graph, it suffices to maximize the log probability:

$$\operatorname*{argmax}_{G} \log Pr(G \mid \mathcal{D}) = \operatorname*{argmax}_{G} \log Pr(\mathcal{D} \mid G) + \log Pr(G).$$

When using the BDeu score, the marginal likelihood decomposes as in Eq. 1. We assume the BDeu score for the remainder of this section.

First, we consider how to update the weights on the edges of a BN graph, to handle a prior on the structure, decomposable or otherwise.

Theorem 1. *Let $Pr_i(.)$ denote a distribution over DAGs having i nodes (i.e., our structure prior).[3] If we label each edge $G_i \xrightarrow{X\mathbf{U}} G_j$ in graph $\mathcal{G}_{\mathrm{bn}}$ with the cost:*

$$\mathsf{score}(X\mathbf{U} \mid \mathcal{D}) - \log \frac{Pr_j(G_j)}{Pr_i(G_i)},$$

then the total cost of a path from the root G_0 to a leaf G_n is

$$\mathsf{score}(G \mid \mathcal{D}) - \log Pr_n(G_n).$$

Hence, assuming a structure prior, the DAG with an optimal score corresponds to a shortest path in the BN graph $\mathcal{G}_{\mathrm{bn}}$, from the root G_0 (top layer) to a leaf G_n (bottom layer). In what follows, we shall assume that our structure prior is a uniform order-modular prior, although general (weighted) order-modular priors can also be accommodated.

We now propose a simple heuristic function for learning an optimal DAG with a uniform order-modular prior. Let $G_i \rightsquigarrow G_n$ indicate that a DAG G_n is reachable from DAG G_i in $\mathcal{G}_{\mathrm{bn}}$. We propose to use the heuristic function $h(G_i) = h_1(G_i) + h_2(G_i)$, which has two components. The first component is:

$$h_1(G_i) = \min_{G_n : G_i \rightsquigarrow G_n} \sum_{X\mathbf{U} \in G_n - G_i} \mathsf{score}(X\mathbf{U} \mid \mathcal{D}) \tag{3}$$

where we sum over families $X\mathbf{U}$ that appear in G_n but not in G_i. This component is looking for the shortest path to the goal, based on the decomposable part of the score, ignoring the prior (i.e., maximizing the marginal likelihood). The second component is:

$$h_2(G_i) = \min_{G_n : G_i \rightsquigarrow G_n} - \log \frac{Pr_n(G_n)}{Pr_i(G_i)} \tag{4}$$

This component is looking for the shortest path to the goal, based on the prior part of the score, but ignoring the data (i.e., maximizing the prior).

Theorem 2. *The heuristic function $h(G_i) = h_1(G_i) + h_2(G_i)$ of Eqs. 3 and 4 is admissible.*

---

[3] $Pr_0(G_0) = 1$ as there is a unique graph over zero nodes.

To use this heuristic function, we must perform two independent optimization problems (for h_1 and h_2). The first is the familiar optimization of a decomposable score; we can employ most any existing structure learning algorithm for decomposable scores, as an oracle, as discussed in the previous section. The second is an optimization of the prior, independently of the data. Next, we show how to both optimize and evaluate this component, for order-modular priors.

Optimizing the Prior. Here, we briefly describe how to solve the component h_2 for a uniform order-modular prior. Again, we want to identify the most likely goal node G_n reachable from G_i, i.e., the DAG G_n with the largest linear extension count. Remember that DAG G_i has i nodes. Since adding any edge to the DAG constrains its possible linear extensions, then the DAG G_n with the largest linear extension count simply adds the remaining $n - i$ nodes independently to DAG G_i. If $\#G_i$ is the linear extension count of DAG G_i, then

$$\#G_n = \#G_i \cdot (i+1) \cdots n$$

is the linear extension count of DAG G_n.[4] Next, we have that:

$$Pr_i(G_i) = \frac{1}{C_i} \cdot \#G_i \quad \text{and} \quad Pr_n(G_n) = \frac{1}{C_n} \cdot \#G_n$$

where C_k is a normalizing constant:

$$C_k = \sum_{G_k} \#G_k = \sum_{G_k} \sum_{\pi \sim G_k} 1 = \sum_{\pi} \sum_{\pi \sim G_k} 1 = \sum_{\pi} 2^{\binom{k}{2}} = k! \cdot 2^{\binom{k}{2}}$$

and where $\pi \sim G_k$ denotes compatibility with an ordering π and a DAG G_k. Thus,

$$\frac{Pr_n(G_n)}{Pr_i(G_i)} = \frac{C_i}{C_n} \frac{\#G_n}{\#G_i} = \frac{C_i}{C_n} \cdot (i+1) \cdots n = 2^{\binom{i}{2} - \binom{n}{2}}$$

Hence, $h_2(G_i) = [\binom{n}{2} - \binom{i}{2}] \cdot \log 2$. We note that for all DAGs G_i in the same layer, the heuristic function $h_2(G_i)$ evaluates to the same value, although this value differs for DAGs in different layers.

Note, that we also need to be able to compute the linear-extension counts $\#G_i$ themselves, which is itself a non-trivial problem (it is #P-complete). We discuss this next.

Counting Linear Extensions. In Sect. 4.1, we highlighted the relationship between uniform order-modular priors and counting linear extensions. We now show that the BN graph itself facilitates the counting of linear extensions, for

---

[4] For each linear extension π of G_i, there are $(i+1)$ places to insert the $(i+1)$-th node, then $(i+2)$ places to insert the next, and so on. Thus, there are $(i+1) \cdots n$ ways to extend a given ordering over i variables to n variables.

many DAGs at once. Subsequently, we shall show that this computation can be embedded in the A* search itself.

Recall that any path in the BN graph $\mathcal{G}_{bn}$, from G_0 to G_i, corresponds to an ordering of variables $\langle X_1, \ldots, X_i \rangle$. In fact, this ordering is a linear extension of the DAG G_i (by construction). Hence, the linear extension count $\#G_i$ of a graph G_i is the number of paths from the root G_0 to G_i, in the BN graph $\mathcal{G}_{bn}$. For example, consider the DAG:

$$\boxed{X_1 \rightarrow X_2 \quad X_3}$$

There are 3 distinct paths in $\mathcal{G}_{bn}$ from the root G_0 to the DAG above, one path for each topological order that the DAG is consistent with. Next, observe that the number of linear extensions of a DAG G_i, (or equivalently, the number of paths that reach G_i), is simply the sum of the linear extensions of the parents of G_i, in the BN graph $\mathcal{G}_{bn}$. For example, our DAG above has 3 linear extensions, and 2 parents in $\mathcal{G}_{bn}$:

$$\boxed{X_1 \rightarrow X_2 \,|\, X_1 \quad X_3}$$

the first with one linear extension, and the second with two. In this way, we can count the linear extensions of DAGs in a BN graph $\mathcal{G}_{bn}$, from top-to-bottom, sharing computations across the different DAGs. A similar algorithm for counting linear extensions is described in, e.g., [24].

Consider how A* navigates the BN graph $\mathcal{G}_{bn}$ during search. If A* expands a node only when all of its parents are expanded, then as described above, we can count the number of linear extensions of a DAG, when it gets expanded.[5] Thus, we can evaluate its prior, and in turn, the function f. It so happens that, we can moderately weaken the heuristic function that we just described, so that A* will in fact expand a node only when all of its parents are expanded.

Theorem 3. *Assuming a uniform order-modular prior, the heuristic function*

$$h(G_i) = h_1(G_i) + h_2'(G_i)$$

allows A* to count the linear extensions of any DAG it expands, where $h_2'(G_i) = -\sum_{k=i+1}^{n} \log k \leq h_2(G_i)$ with components h_1 and h_2 coming from Eqs. 3 and 4.

A proof appears in the Appendix.

4.3   A* Search

Algorithm 1 provides pseudo-code for A* search using a uniform order-modular prior. Note that this pseudo-code deviates slightly from the standard A* search, as the linear extension counts $\#G$ are computed incrementally during the search.

---

[5] In particular, every time that we expand a node G, we can increment each of its children's linear extension counts by $\#G$. Once we have expanded every parent of a child, the child's linear extension count is correct.

Algorithm 1. A* search for learning an optimal BN with a uniform order-modular prior.

Data: a dataset $\mathcal{D}$ over variables $\mathbf{X}$
Result: an optimal BN maximizing $Pr(G \mid \mathcal{D})$
begin
    $H \leftarrow$ min-heap with only $(G_0, f(G_0), 1)$, where 1 is the number of linear
    extensions of G_0; and the heap is ordered by f
    while $H \neq \emptyset$ **do**
        extract the minimum item (G, f, l) from H
        if $V(G) = \mathbf{X}$ **then** return G
        foreach G' obtained by adding a leaf to G **do**
            if G' is not in H **then**
                insert into H: $(G', \mathsf{score}(G'|\mathcal{D}) - \log l + h(G'), l)$
            else
                let (G', f', l') be in H, decrease f' by $\log \frac{l'+l}{l'}$, increase l' by l;
                and reheapify
            end
        end
    end
end

Theorem 4. *Algorithm 1 learns an optimal Bayesian network with a uniform order-modular prior.*

A proof appears in the Appendix.

Finally, we note that Theorem 1 and the heuristic functions of Eq. 3 and 4 were proposed for order-modular priors. In principle, the shortest-path formulation, and the heuristic function that we proposed, can support a much broader class of non-decomposable priors. However, one must be able to optimize the probability of a graph, as in the component h_2 of the heuristic function that we proposed, in Eq. 4. If we had access to some oracle that can solve this component, then we would in principle have the pieces that are sufficient to perform A* search over the DAG graph $\mathcal{G}_{\mathrm{bn}}$, using the corresponding prior.

4.4 Experiments

We evaluate our A* search approach to learning optimal Bayesian networks with real-world datasets, assuming a uniform order-modular prior. In Table 1, we find that our approach can scale up to 17 variables on real-world datasets (i.e., the letter and voting datasets). We also note that with more data, and with more of the probability mass concentrated on fewer DAGs, traversing the BN graph with A* search appears to become more efficient. In particular, consider the time spent in A* search (T_{A*}), and the number of nodes generated (gen.), in the datasets adult and wine, which both have 14 variables. Similarly, consider the datasets letter and voting, which both have 17 variables. Moreover, consider

Table 1. The performance of A* search on the BN graph when learning with the uniform order-modular prior: (1) The time T_h to compute the heuristic function. (2) The time T_{A*} to traverse the BN graph with A* (in seconds) (3) The total time $t = T_h + T_{A*}$ spent in A* (4) The number of generated nodes. (5) The number of expanded nodes. (6) The number of re-expanded nodes (in partial-expansion A*). An $\times_m$ corresponds to an out-of-memory (64 GB).

Benchmark	n	N	T_h	T_{A*}	t	Gen.	Exp.	Re-exp.
Adult	14	30,162	1.03	0.26	1.29	106,832	12,620	33
Wine	14	435	0.74	6.08	6.82	1,559,900	244,694	57,259
Nltcs	16	16,181	7.21	1.17	8.38	386,363	41,125	1
Letter	17	20,000	29.42	3.79	32.20	360,899	37,034	16
Voting	17	435	5.28	56.59	61.89	10,540,132	1,961,602	396,084
Zoo	17	101				$\times_m$		

dataset zoo, also over 17 variables, which was a very small dataset, containing only 101 instances. Here, A* search exhausted the 64 GB of memory that it was allowed. We remark that, to our knowledge, ours is the first system for finding optimal DAGs using order-modular priors.[6]

In Fig. 3, we consider a simple example, highlighting the effect that a uniform order-modular prior can have on the structure we learn. In Fig. 3(a), we have the classical asia network, which we used to simulate datasets of different sizes. First, we simulated a small dataset of size 2^7 and learned two networks, one with a prior, Fig. 3(b), and one without a prior, Fig. 3(c). Ignoring node A, the two networks are Markov equivalent. However, including node A, their linear extension counts are very different: 96 for network Fig. 3(b) but only 3 for network Fig. 3(c). This difference can explain why variable A is disconnected in Fig. 3(b), as a disconnected node non-trivially increases the linear extension count (and hence, the weight of the prior). In Fig. 3(d), both cases (with and without the prior) learned precisely the same network when we raised the size of the dataset to 2^{14} (this DAG has 150 linear extensions). This network is Markov equivalent to the ground truth network that generated the data.

5   Enumerating the k-Best Structures

We will next show how we can use our proposed framework for learning Bayesian networks, using BN graphs, to enumerate the k-best Bayesian network structures.

---

[6] There are systems available for (a) finding optimal DAGs using structure-modular priors, (b) for Bayesian model averaging using order-modular priors, and (c) for jointly optimizing over orders and DAGs, using order-modular priors. These tasks are all discussed in [18], which further states that finding optimal DAGs with order-modular priors is a more challenging problem (where we maximize over DAGs, but sum over orders).

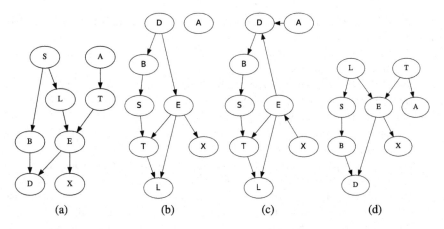

Fig. 3. A network `asia` (a), and networks learned with dataset size 2^7 with a prior (b), without a prior (c), and a network learned with dataset size 2^{14} (d).

Enumerating the k-best Bayesian network structures is particularly simple to do when we perform A* search on the BN graph $\mathcal{G}_{\mathrm{bn}}$. In particular, the k-th best DAG can be obtained by finding the goal node G_n that is the k-th closest to the root G_0. We can thus enumerate the k-best DAGs by simply continuing the A* search, rather than stopping when we reach the first goal node; see [12], for more on using A* search for k-best enumeration.[7,8]

We also note that if our heuristic is perfect, then we can enumerate all DAGs with an optimal score, relatively efficiently. In particular, A* will *only* expand nodes that lead to an optimal DAG, as long as a DAG with an optimal score remains (typically, they are all Markov equivalent). However, once we have exhausted all optimal DAGs, our heuristic is no longer perfect; cf. [12]. Given a DAG G, our heuristic is still admissible, as it still lower-bounds the cost of the possible extensions to a goal node G_n. That is, it may just report a cost for a goal node that was already enumerated (and hence has a lower cost). We can thus continue to employ the same heuristic in A* search, to enumerate the remaining k-best DAGs.

---

[7] We remark, however, that [12] is more specifically concerned with the enumeration of the k-shortest paths. Since we are interested in enumerating the k-closest goal nodes, we remark that some, but not all, of their theoretical analyses applies to our problem. In particular, each distinct goal node in the BN graph may have many paths that can reach it. Hence, once we obtain one goal node, many more shortest-paths may be needed to obtain the next closest (and distinct) goal node. Moreover, we do not need to differentiate between two different paths to the same goal node, as in [12].

[8] We remark on another distinction between finding a single optimal DAG, versus enumerating the k-best DAGs. In particular, there are techniques that can guarantee that certain families will not appear in an optimal DAG, which can greatly simplify the learning problem [8,11,30,31]. However, such families may still appear in a k-th best DAG, and hence, these techniques may not be directly applicable.

We further note a distinction between the BN graph and the order graph. In the BN graph, DAGs are represented explicitly, whereas in an order graph, DAGs are implicit. In particular, each node $\mathbf{Y}$ in the order graph represents just a single optimal DAG over the variables $\mathbf{Y}$. Hence, the k-th best DAG may not be immediately recoverable. That is, we may not be able to obtain the k-th best DAG starting from an optimal sub-DAG—we can only guarantee that we obtain a single optimal DAG. While it is possible to augment the order graph to find the k-th best DAG, as in [32], this is not as effective as searching the BN graph, as we shall soon see.

Finally, we consider another approach to enumerating the k-best DAGs in our experiments, based on integer linear programming (ILP) [9]. Basically, once an optimal DAG is obtained from an ILP, a new ILP can be obtained, whose optimal solution corresponds to the next-best DAG. In particular, we assert additional constraints that exclude the optimal DAG that we found originally. This process can be repeated to enumerate the k-best DAGs.

Next, we empirically compare the DP and ILP approaches, with our proposal, based on A* search in the BN graph.

5.1   Experiments

We compare our approach, which is based on A* search, with two other recently proposed k-best structure learning algorithms: (1) KBEST,[9] which is based on dynamic programming (DP) [32], and (2) GOBNILP,[10] which is based on integer linear programming (ILP) [7].

For each approach, we enumerate the 10-best, 100-best and 1,000-best BNs, over a variety of real-world datasets. We impose a 7,200 second limit on running time. To analyze memory usage, we incrementally increased the amount of memory available to each system (from 1 GB, 2 GB, 4 GB, 8 GB, 16 GB, and up to 64 GB), and recorded the smallest limit that allowed each system to finish.

Table 2 summarizes our results for A* search on the BN graph, and for the DP-based approach of KBEST. We omit the results for the ILP-based approach of GOBNILP, which ran out-of-memory (given 64 GB) for all instances, except for the case of 10-best networks on the wine dataset, which took 2,707.13 s and under 8 GB of memory.[11]

We observe a few trends. First, A* search on the BN graph can be over *three orders of magnitude* more efficient than KBEST, at enumerating the k-best BNs. For example, when we enumerate the 100-best BNs on the voting dataset, A* search is over 4,323 times faster. Next, we observe that A* search is consistently more efficient than KBEST at enumerating the k-best networks

---

[9] At http://www.cs.iastate.edu/~jtian/Software/UAI-10/KBest.htm.

[10] At http://www.cs.york.ac.uk/aig/sw/gobnilp/.

[11] Note that GOBNILP is known to be more effective in other regimes, for example, where we can constrain the number of parents that a node can have [21,34]. However, for our experiments here, we consider the more general case, where we do not assume such a constraint.

Table 2. A comparison of the time t (in seconds) and memory m (in GBs) used by BN graph and KBEST. An $\times_m$ corresponds to an out-of-memory (64 GB), and an $\times_t$ corresponds to an out-of-time (7,200 s). n denotes the number of variables in the dataset, and N denotes the size of the dataset.

| Benchmark | | | 10-best | | | | 100-best | | | | 1,000-best | | | |
| | | | BN graph | | KBEST | | BN graph | | KBEST | | BN graph | | KBEST | |
Name	n	N	t	m	t	m	t	m	t	m	t	m	t	m
Wine	14	435	0.16	1	5.24	1	0.24	1	162.69	1	0.73	1	4,415.98	4
Nlts	16	16,181	2.84	1	18.84	1	4.01	1	787.52	1	5.67	1	$\times_t$	
Letter	17	20,000	13.38	1	42.16	1	15.85	1	1,849.29	2	19.40	1	$\times_t$	
Voting	17	435	0.39	1	59.29	1	0.58	1	2,507.72	2	2.85	1	$\times_t$	
Zoo	17	101	4.45	1	58.25	1	4.97	1	2,236.13	2	7.71	1	$\times_t$	
Statlog	19	752	58.82	4	291.88	1	76.17	4	$\times_t$		78.47	4	$\times_t$	
Hepatitis	20	126	14.95	4	675.34	2	29.53	8	$\times_t$		66.96	8	$\times_t$	
Image	20	2,310	344.54	8	480.03	2	344.54	8	$\times_t$		344.61	8	$\times_t$	
Imports	22	205	3,013.97	32	2,646.41	8	3,167.11	32	$\times_t$		3,167.88	32	$\times_t$	
Parkinsons	23	195	3,728.96	64	6,350.58	16	3,730.56	64	$\times_t$		4,125.30	64	$\times_t$	
Sensors	25	5,456	$\times_m$		$\times_t$		$\times_m$		$\times_t$		$\times_m$		$\times_t$	

Table 3. The time T_h to compute the heuristic function and the time T_{A*} to traverse the BN graph with A* (in seconds).

| Benchmark | | 10-best | | 100-best | | 1,000-best | |
Name	n	T_h	T_{A*}	T_h	T_{A*}	T_h	T_{A*}
Wine	14	0.14	0.02	0.14	0.10	0.19	0.55
Nltcs	16	2.83	0.01	3.96	0.05	5.27	0.39
Letter	17	13.36	0.02	15.72	0.13	18.28	1.12
Voting	17	0.36	0.02	0.38	0.19	1.05	1.80
Zoo	17	4.44	0.01	4.93	0.04	7.35	0.35
Statlog	19	58.79	0.03	75.99	0.18	77.58	0.89
Hepatitis	20	14.90	0.05	29.20	0.33	64.56	2.40
Image	20	344.53	0.01	344.53	0.01	344.53	0.08
Imports	22	3,013.39	0.59	3,166.36	0.75	3,166.63	1.26
Parkinsons	23	3,728.31	0.65	3,729.33	1.23	4,117.36	7.94

(except for dataset imports for $k = 10$). In general, our approach scales to larger networks (with more variables), and to larger values of k. In fact, KBEST appears to scale super-linearly with k, but A* search appears to scale *sub-linearly* with respect to k. These differences are due in part to: (1) the more exhaustive nature of dynamic programming (DP) (we need to maintain all of the partial solutions that can potentially be completed to a k-th best solution), and (2) the more incremental nature of A* (the next best solutions are likely to be in the priority queue already). Finally, we see that the memory usage of these two approaches is comparable, although memory usage by A* search appears to be more memory efficient as we increase the number k of networks that we enumerate.

Table 4. (1) The number of generated nodes. (2) The number of expanded nodes. (3) The number of re-expanded nodes (in partial-expansion A*).

Benchmark		10-best			100-best			1,000-best		
Name	n	Gen	Exp	Re-exp	Gen	Exp	Re-exp	Gen	Exp	Re-exp
Wine	14	13,156	1,948	0	50,754	8,110	36	254,981	40,992	957
Nltcs	16	1,847	202	0	19,542	2,145	0	173,726	19,784	0
Letter	17	2,613	285	0	19,795	2,174	0	164,794	18,149	0
Voting	17	13,646	1,884	0	89,153	13,836	246	727,944	118,779	4,141
Zoo	17	1,848	205	0	9,994	1,165	0	89,262	10,808	0
Statlog	19	3,603	410	0	30,517	3,589	0	223,069	26,941	68
Hepatitis	20	16,854	2,165	0	114,054	15,897	2	757,727	111,542	816
Image	20	318	43	0	2,546	397	0	31,974	4,903	0
Imports	22	2,217	251	0	20,781	2,416	0	130,426	15,923	84
Parkinsons	23	893	104	0	14,054	1,679	494	134,745	16,197	0

Table 5. The number of times the black-box is invoked to evaluate the heuristic function.

Benchmark	n	10-best	100-best	1,000-best
Wine	14	896	896	1,067
Nltcs	16	136	402	683
Letter	17	182	472	744
Voting	17	720	1,867	4,779
Zoo	17	289	518	1,679
Statlog	19	230	1,058	1,711
Hepatitis	20	2,235	9,037	26,499
Image	20	124	130	142
Imports	22	234	654	694
Parkinsons	23	155	494	2,065

To gain more insight about the computational nature (and bottlenecks) of A* search on the BN graph, consider Table 3, which looks at how much time T_h that was spent on evaluating the heuristic function, versus the time T_{A*} that was spent in navigating the BN graph (where $t = T_h + T_{A*}$, with the total time t corresponding to those reported in Table 2). Table 4 further reports the number of nodes generated (the number of nodes inserted into the open list) and expanded by A* search. First, we observe that the vast majority of the time spent in search is spent in evaluating the heuristic function. This is expected, as evaluating our black-box heuristic function is relatively expensive. Next, we observe that the number of nodes expanded is relatively small, which suggests that our black-box heuristic is indeed powerful enough to efficiently navigate the large search space of the BN graph. We also remark again that due to the caching of heuristic values (which we discussed earlier), the number of times that our black-box is invoked, can be much smaller than the number of times that a node is generated. This is illustrated in Table 5.

6   Conclusion

Underlying nearly all score-based methods for learning Bayesian networks from data, is the property of score decomposability. This has been true, since the first Bayesian network learning algorithms were proposed, over two decades ago. While the property of score decomposability has been fruitfully exploited over this time, there is a notable void in the structure learning landscape, in the support of learning with non-decomposable scores. This includes a general lack of support for the integration of more expressive, but non-decomposable forms of prior knowledge.

In this paper, we take a step towards a more general framework for Bayesian network structure learning that targets this void. We proposed a new search space, called the BN graph, which explicates all Bayesian network structures. We proposed to navigate this tremendously large search space, with the assistance of a (nearly) omniscient oracle—any state-of-the-art system for Bayesian network structure learning can be used as this oracle. Using heuristic search methods, such as A* search, we showed how this framework can be used to find optimal Bayesian network structures, using non-decomposable scores (even when our oracle relies on decomposable scores). To our knowledge, ours is the first system for finding optimal DAGs using order-modular priors, in particular. Further, we showed that enumerating the k-best DAGs is very simple on the BN graph, where empirically, we observed three orders of magnitude improvement, compared to existing approaches.

Acknowledgments. We thank Joseph Barker, Zhaoxing Bu, and Ethan Schreiber for helpful comments and discussions. We also thank James Cussens and Brandon Malone for their comments on an earlier version of this paper. This work was supported in part by ONR grant #N00014-12-1-0423 and NSF grant #IIS-1514253.

A   Proofs for Sect. 4.2

Proof (Theorem 1). The total cost of a path from the root G_0 to a leaf G_n is:

$$\sum_{G_i \to G_j} \mathsf{score}(X_j \mathbf{U}_j \mid \mathcal{D}) - \log \frac{Pr_j(G_j)}{Pr_i(G_i)}$$
$$= \sum_{G_i \to G_j} \mathsf{score}(X_j \mathbf{U}_j \mid \mathcal{D}) - \log Pr_j(G_j) + \log Pr_i(G_i)$$

$$= \mathsf{score}(X_1\mathbf{U}_1 \mid \mathcal{D}) - \log Pr_1(G_1) + \log Pr_0(G_0)$$
$$+ \mathsf{score}(X_2\mathbf{U}_2 \mid \mathcal{D}) - \log Pr_2(G_2) + \log Pr_1(G_1) + \dots$$
$$+ \mathsf{score}(X_n\mathbf{U}_n \mid \mathcal{D}) - \log Pr_n(G_n) + \log Pr_{n-1}(G_{n-1})$$
$$= \mathsf{score}(G \mid \mathcal{D}) - \log Pr_n(G_n)$$

as desired. $\square$

Proof (Theorem 2).

$$h(G_i) = \min_{G_n:G_i \rightsquigarrow G_n} \sum_{X\mathbf{U} \in G_n - G_i} \mathsf{score}(X\mathbf{U} \mid \mathcal{D}) + \min_{G_n:G_i \rightsquigarrow G_n} - \log \frac{Pr_n(G_n)}{Pr_i(G_i)}$$

$$\leq \min_{G_n:G_i \rightsquigarrow G_n} \left(\sum_{X\mathbf{U} \in G_n - G_i} \mathsf{score}(X\mathbf{U} \mid \mathcal{D}) - \log \frac{Pr_n(G_n)}{Pr_i(G_i)} \right)$$

$$= \min_{G_n:G_i \rightsquigarrow G_n} g(G_n) - g(G_i)$$

Since heuristic function h lower-bounds the true cost of to a goal node, it is admissible. $\square$

Below we consider the correctness of Algorithm 1.

Lemma 1. *In Algorithm 1:*

1. *all G_i that generate G_{i+1} are extracted from H before G_{i+1} is extracted;*
2. *when $(G_{i+1}, f_{i+1}, l_{i+1})$ is extracted from H,*

$$f_{i+1} = \mathsf{score}(G_{i+1}|\mathcal{D}) - \log \#G_{i+1} + h(G_{i+1}),$$

and l_{i+1} is the number of linear extensions of G_{i+1}, i.e., $\#G_{i+1}$.

where $h(G_i) = h_1(G_i) - \sum_{j=i+1}^{n} \log j$.

Proof. Consider a minimum item $(G_{i+1}, f_{i+1}, l_{i+1})$ extracted from H. Below we show by induction that (1) all G_i such that G_i generates G_{i+1} are extracted from H before G_{i+1} (2) $f_{i+1} = \mathsf{score}(G_{i+1}|\mathcal{D}) - \log \#G_{i+1} + h(G_{i+1})$, and l_{i+1} is the number of linear extensions of G_{i+1}, which is also the number of paths from G_0 to G_{i+1}.

For $i = 0$, clearly (1) and (2) are true. Assume (1) and (2) are true for all (G_i, f_i, l_i). Then when $(G_{i+1}, f_{i+1}, l_{i+1})$ is extracted, l_{i+1} is the number of paths from G_0 to G_{i+1} that pass the G_i extracted from H before G_{i+1}. To see this, note that l is the number of path from G_0 to G_i. Moreover, since l_{i+1} is the number paths from G_0 to G_{i+1} that pass the G_i extracted from H before G_{i+1}, when $(G_{i+1}, f_{i+1}, l_{i+1})$ is in H,

$$f_{i+1} = \mathsf{score}(G_{i+1}|\mathcal{D}) - \log \sum_{G_i \prec G_{i+1}} N(G_0 \to \dots \to G_i \to G_{i+1}) + h(G_{i+1}),$$

where $G_i \prec G_{i+1}$ denotes that G_i is extracted before G_{i+1}, and $N(G_0 \rightarrow \ldots \rightarrow G_i \rightarrow G_{i+1})$ denotes the number of paths $G_0 \rightarrow \ldots \rightarrow G_i \rightarrow G_{i+1}$. Note that f_{i+1} decreases as more G_i are extracted.

Consider when (G_i, f_i, l_i) and $(G_{i+1}, f_{i+1}, l_{i+1})$ are both in H. Below we show that all G_i that generates G_{i+1} are extracted from H before G_{i+1}. Consider

$$f_i = \mathsf{score}(G_i|\mathcal{D})$$

$$- \log \sum_{G_{i-1} \prec G_i} N(G_0 \rightarrow \ldots \rightarrow G_{i-1} \rightarrow G_i) + h_1(G_i) - \sum_{j=i+1}^{n} \log j$$

$$f_{i+1} = \mathsf{score}(G_{i+1}|\mathcal{D})$$

$$- \log \sum_{G_i' \prec G_{i+1}} N(G_0 \rightarrow \ldots \rightarrow G_i' \rightarrow G_{i+1}) + h_1(G_{i+1}) - \sum_{j=i+2}^{n} \log j$$

We simply need to show $f_{i+1} > f_i$. Since h_1 is a consistent heuristic function for learning with score, $\mathsf{score}(G_{i+1}|\mathcal{D}) + h_1(G_{i+1}) \geq \mathsf{score}(G_i|\mathcal{D}) + h_1(G_i)$. Then we only need to show

$$(i+1) \sum_{G_{i-1} \prec G_i} N(G_0 \rightarrow \ldots \rightarrow G_{i-1} \rightarrow G_i)$$

$$> \sum_{G_i' \prec G_{i+1}} N(G_0 \rightarrow \ldots \rightarrow G_i' \rightarrow G_{i+1})$$

First, for any pair of DAGs G_i and G_i' that can generate a DAG G_{i+1}, there exists a unique DAG G_{i-1} that can generate both G_i and G_i'. For each G_i' on the right-hand side, there thus exists a corresponding (and unique) G_{i-1} on the left-hand side that can generate both G_i' and G_i. Further, since G_i' was expanded, G_{i-1} must also have been expanded (by induction). For each such G_{i-1}, if G_i' has a linear extension count of L, then G_{i-1} must have at least a linear extension count of L/i, and hence the corresponding $N(G_0 \rightarrow \ldots \rightarrow G_{i-1} \rightarrow G_i)$ is at least L/i. On the left-hand side, we the corresponding term is thus at least $(i+1) \cdot L/i > L$. Since this holds for each element of the summation on the right-hand side, the above inequality holds.

Since all G_i that generates G_{i+1} are extracted from H before G_{i+1}, as a result, $f_{i+1} = \mathsf{score}(G_{i+1}|\mathcal{D}) - \log \#G_{i+1} + h(G_{i+1})$, and l_{i+1} is the number of all paths from G_0 to G_{i+1}.

Proof (of Theorem 4). To see the correctness of the algorithm, simply note that by Lemma 1, when $(G_{i+1}, f_{i+1}, l_{i+1})$ is extracted from H, i.e. the open list, $f_{i+1} = f(G_{i+1})$.

Proof (of Theorem 3). By Lemma 1, Algorithm 1 can count the number of linear extensions.

References

1. Bache, K., Lichman, M.: UCI Machine Learning Repository (2013). http://archive.ics.uci.edu/ml
2. Bouckaert, R.R.: Probalistic network construction using the minimum description length principle. In: Proceedings of the European Conference on Symbolic and Quantitative Approaches to Reasoning and Uncertainty (ECSQARU), pp. 41–48 (1993)
3. Brightwell, G., Winkler, P.: Counting linear extensions. Order **8**(3), 225–242 (1991)
4. Buntine, W.: Theory refinement on Bayesian networks. In: Proceedings of the Seventh Conference on Uncertainty in Artificial Intelligence, pp. 52–60 (1991)
5. Chickering, D., Geiger, D., Heckerman, D.: Learning Bayesian networks: search methods and experimental results. In: Proceedings of the Fifth International Workshop on Artificial Intelligence and Statistics (AISTATS), pp. 112–128 (1995)
6. Cooper, G.F., Herskovits, E.: A Bayesian method for the induction of probabilistic networks from data. Mach. Learn. **9**(4), 309–347 (1992)
7. Cussens, J.: Bayesian network learning with cutting planes. In: Proceedings of the Twenty-Seventh Conference on Uncertainty in Artificial Intelligence, pp. 153–160 (2011)
8. Cussens, J.: An upper bound for BDeu local scores. In: European Conference on Artificial Intelligence Workshop on Algorithmic Issues for Inference in Graphical Models (2012)
9. Cussens, J., Bartlett, M., Jones, E.M., Sheehan, N.A.: Maximum likelihood pedigree reconstruction using integer linear programming. Genet. Epidemiol. **37**(1), 69–83 (2013)
10. Darwiche, A.: Modeling and Reasoning with Bayesian Networks. Cambridge University Press, New York (2009)
11. De Campos, C.P., Ji, Q.: Efficient structure learning of Bayesian networks using constraints. J. Mach. Learn. Res. **12**, 663–689 (2011)
12. Dechter, R., Flerova, N., Marinescu, R.: Search algorithms for m best solutions for graphical models. In: Proceedings of the Twenty-Sixth Conference on Artificial Intelligence (2012)
13. Felner, A., Goldenberg, M., Sharon, G., Stern, R., Beja, T., Sturtevant, N.R., Schaeffer, J., Holte, R.: Partial-expansion A* with selective node generation. In: Proceedings of the Twenty-Sixth Conference on Artificial Intelligence (2012)
14. Friedman, N., Koller, D.: Being Bayesian about network structure. In: Proceedings of the Sixteenth Conference on Uncertainty in Artificial Intelligence, pp. 201–210 (2000)
15. Hart, P.E., Nilsson, N.J., Raphael, B.: A formal basis for the heuristic determination of minimum cost paths. IEEE Trans. Syst. Sci. Cybern. **4**(2), 100–107 (1968)
16. Heckerman, D., Geiger, D., Chickering, D.M.: Learning Bayesian networks: the combination of knowledge and statistical data. Mach. Learn. **20**(3), 197–243 (1995)
17. Jaakkola, T., Sontag, D., Globerson, A., Meila, M.: Learning Bayesian network structure using LP relaxations. In: Proceedings of the Thirteen International Conference on Artificial Intelligence and Statistics, pp. 358–365 (2010)
18. Koivisto, M., Sood, K.: Exact Bayesian structure discovery in Bayesian networks. J. Mach. Learn. Res. **5**, 549–573 (2004)
19. Koller, D., Friedman, N.: Probabilistic Graphical Models: Principles and Techniques. The MIT Press, Cambridge (2009)

20. Lam, W., Bacchus, F.: Learning Bayesian belief networks: an approach based on the MDL principle. Comput. Intell. **10**, 269–294 (1994)
21. Malone, B., Kangas, K., Järvisalo, M., Koivisto, M., Myllymäki, P.: Predicting the hardness of learning Bayesian networks. In: Proceedings of the Twenty-Eighth Conference on Artificial Intelligence (2014)
22. Malone, B., Yuan, C., Hansen, E.: Memory-efficient dynamic programming for learning optimal Bayesian networks. In: Proceedings of the Twenty-Fifth International Joint Conference on Artificial Intelligence (2011)
23. Murphy, K.P.: Machine Learning: A Probabilistic Perspective. MIT Press, Cambridge (2012)
24. Niinimäki, T.M., Koivisto, M.: Annealed importance sampling for structure learning in Bayesian networks. In: Proceedings of the 23rd International Joint Conference on Artificial Intelligence (IJCAI) (2013)
25. Oates, C.J., Smith, J.Q., Mukherjee, S., Cussens, J.: Exact estimation of multiple directed acyclic graphs. Stat. Comput. 1–15 (2015). http://link.springer.com/journal/11222/onlineFirst/page/2
26. Russell, S.J., Norvig, P.: Artificial Intelligence - A Modern Approach. Pearson Education, London (2010)
27. Silander, T., Myllymäki, P.: A simple approach for finding the globally optimal Bayesian network structure. In: Proceedings of the Twenty-Second Conference on Uncertainty in Artificial Intelligence, pp. 445–452 (2006)
28. Singh, A.P., Moore, A.W.: Finding optimal Bayesian networks by dynamic programming. Technical report, CMU-CALD-050106 (2005)
29. Suzuki, J.: A construction of Bayesian networks from databases based on an MDL principle. In: Proceedings of the Ninth Annual Conference on Uncertainty in Artificial Intelligence (UAI), pp. 266–273 (1993)
30. Teyssier, M., Koller, D.: Ordering-based search: a simple and effective algorithm for learning Bayesian networks. In: Proceedings of the Twenty-first Conference on Uncertainty in Artificial Intelligence, pp. 584–590 (2005)
31. Tian, J.: A branch-and-bound algorithm for mdl learning Bayesian networks. In: Proceedings of the Sixteenth Conference on Uncertainty in Artificial Intelligence, pp. 580–588 (2000)
32. Tian, J., He, R., Ram, L.: Bayesian model averaging using the k-best Bayesian network structures. In: Proceedings of the Twenty-Six Conference on Uncertainty in Artificial Intelligence, pp. 589–597 (2010)
33. Yoshizumi, T., Miura, T., Ishida, T.: A* with partial expansion for large branching factor problems. In: Proceedings of the Seventeenth International Joint Conference on Artificial Intelligence (2000)
34. Yuan, C., Malone, B.: Learning optimal Bayesian networks: a shortest path perspective. J. Artif. Intell. Res. **48**, 23–65 (2013)
35. Yuan, C., Malone, B., Wu, X.: Learning optimal Bayesian networks using A* search. In: Proceedings of the Twenty-Second International Joint Conference on Artificial Intelligence, pp. 2186–2191 (2011)

Combinatorial Results on Directed Hypergraphs for the SAT Problem

Cornelius Croitoru[1] and Madalina Croitoru[2][(✉)]

[1] University Al. I. Cuza, Iasi, Romania
[2] LIRMM, University Montpellier, Montpellier, France
Croitoru@lirmm.fr

Abstract. Directed hypergraphs have already been shown to unveil several combinatorial inspired results for the SAT problem. In this paper we approach the SAT problem by searching a transversal of the directed hypergraphs associated to its instance. We introduce some particular clause orderings and study their influence on the backtrack process, exhibiting a new subclass of CNF for which SAT is polynomial. Based on unit resolution and a novel dichotomous search, a new DPLL-like algorithm and a renaming-based combinatorial approach are proposed. We then investigate the study of weak transversals in this setting and reveal a new degree of a CNF formula unsatisfiability and a structural result about unsatisfiable formulae.

1 Introduction

In this paper we consider a representation based on directed hypergraphs of the SAT problem. Directed hypergraphs are generalizations of directed graphs, introduced in [15] to represent deduction properties in data bases as paths in hypergraphs, and developed in several papers (see [1,7]).

The main contribution of the paper is showing how a transversal based directed hypergraph representation of the SAT problem opens the way to new methods and ideas, otherwise not evident without a hypergraph representation. The advantage of this representation is the uniform addressing of assignments and clauses. Our aim is to show the plethora of interesting combinatorial results this formalisation opens up (deepening the results in [6,7]). These contributions will take the form of a new polynomial class of SAT instances (made evident due to a clause ordering natural in the hypergraph setting); a new variable elimination based algorithm that could be used for further optimisations of SAT solvers and a new unsatisfiability measure focusing on problematic variables (and not problematic clauses as traditionally considered in the field).

A *directed hypergraph* is a pair $H = (V, \mathcal{C})$, where V is a finite set of *vertices* and $\mathcal{C} = (C_i; i \in I)$ is a *family of directed edges*; the index set I is a finite set (possibly empty). Each directed edge is a pair $C = (C^+, C^-)$, where $C^+, C^- \subseteq V$.

The propositional logic satisfiability problem consists of finding a truth assignment for the propositional variables of CNF formula F, represented as a

© Springer International Publishing Switzerland 2015
M. Croitoru et al. (Eds.): GKR 2015, LNAI 9501, pp. 72–88, 2015.
DOI: 10.1007/978-3-319-28702-7_5

(multi)set of clauses $\mathcal{C}$ (each clause being a disjunction of propositional variables or their negations), such that each clause evaluates to *true* (F is satisfiable), or proving that such truth assignment does not exist (F is unsatisfiable). SAT, the corresponding decision problem, has as instance a finite set of propositional variables V and the formula $F = \mathcal{C}$ over V and asks if F is satisfiable. Clearly, $H_F = (V, \mathcal{C})$ is a directed hypergraph if each clause $C \in \mathcal{C}$ is viewed as the pair (C^+, C^-), where C^+ is the set of all non-negated variables in C and C^- is the set of all negated variables in C.

Example. Let F be the following propositional sentence in conjunctive normal form (CNF): $\mathcal{C} = (v_1 \vee v_2 \vee v_3) \wedge (\neg v_1 \vee \neg v_2 \vee v_3) \wedge (\neg v_1 \vee v_2 \vee v_3) \wedge (v_2 \vee \neg v_2 \vee v_5) \wedge (v_1 \vee v_5 \vee \neg v_1) \wedge (v_3 \vee v_4 \vee v_5)$, where v_i are propositional variables. The clausal representation of F as a directed hypergraph on $V = \{v_1, \ldots, v_5\}$ is $H_F = (V, (C_1, \ldots, C_6))$, where $C_1 = (\{v_1, v_2, v_3\}, \emptyset)$, $C_2 = (\{v_3\}, \{v_1, v_2\})$, $C_3 = (\{v_2, v_3\}, \{v_1\})$, $C_4 = (\{v_2, v_5\}, \{v_2\})$, $C_5 = (\{v_1, v_5\}, \{v_1\})$, $C_6 = (\emptyset, \{v_3, v_4, v_5\})$. This directed hypergraph can be visualised as in Fig. 1 bellow:

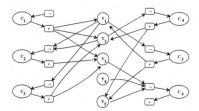

Fig. 1. Directed hypergraph of CNF formula

Please note the "dangling" directed edges C_1 and C_6. Also, the tautological clauses C_4 and C_5 have $C_i^+ \cap C_i^- \neq \emptyset$, which could be omitted in the SAT problem.

Motivated by the results and algorithms generalizing the similar one for directed graphs, in [6,7] it is shown that F is satisfiable if and only if in the directed hypergraph H_F there is a 0-cardinality cutset: a (weak) partition (V^+, V^-) of V ($V^+ \cup V^- = V$ and $V^+ \cap V^- = \emptyset$) such that there is **no** clause $C \in \mathcal{C}$ with $C^- \subseteq V^+$ and $C^+ \subseteq V^-$. This is equivalent to the fact that there is a pair (A^+, A^-) of subsets of V such that $A^+ \cap A^- = \emptyset$ and, for each clause $C = (C^+, C^-) \in \mathcal{C}$, we have $A^+ \cap C^+ \neq \emptyset$ or $A^- \cap C^- \neq \emptyset$. Setting the variable from A^+ to *true* and the variables from A^- to *false*, we obtain a (partial) satisfying assignment for F. Conversely, for each satisfying assignment of F, we can construct a pair (A^+, A^-) with the above properties. It follows that we could approach the SAT problem by searching a transversal of the directed hypergraphs associated to its instances.

Let us precise some notations associated to the subsets pairs of a finite set V. The ***support*** of $C = (C^+, C^-) \in \mathcal{C}$ is $V(C) = C^+ \cup C^- \subseteq V$. The pair $(\emptyset, \emptyset)$ is denoted $2\emptyset$. The intersection and union of two pairs sets $C_1 = (C_1^+, C_1^-)$ and $C_2 = (C_2^+, C_2^-)$ are defined componentwise: $C_1 \sqcap C_2 = (C_1^+ \cap C_2^+, C_1^- \cap C_2^-)$ and

$C_1 \sqcup C_2 = (C_1^+ \cup C_2^+, C_1^- \cup C_2^-)$. Also we denote $C_1 \setminus C_2 = (C_1^+ \setminus C_2^+, C_1^- \setminus C_2^-)$. If $C_1 \sqcap C_2 = C_1$, then $C_1 \sqsubseteq C_2$. $C_1 = C_2$ if and only if $C_1 \sqsubseteq C_2$ and $C_2 \sqsubseteq C_1$.

A ***transversal*** of $H = (V, \mathcal{C})$ is a pair $T = (T^+, T^-)$ such that $T^+, T^- \subseteq V$, $T^+ \cap T^- = \emptyset$ and $T \sqcap C \neq 2\emptyset$, for each $C \in \mathcal{C}$. Note that if $\mathcal{C} = \emptyset$, then any pair $T = (T^+, T^-)$ of disjoint subsets of V is a transversal of H.

We denote by $Tr(H)$ the ***directed hypergraph of all transversals*** of H. By the above discussion, the SAT problem can be reformulated as:

SAT: *Given a directed hypergraph $H = (V, \mathcal{C})$, is $Tr(H)$ nonempty ?*

In the next section we study the SAT problem in this setting, which gives the possibility that clauses and assignments are uniformly addressed. First, we study the impact of the order in which the clauses are considered in a SAT search procedure. We introduce some particular clause orderings and exhibit their influence on the backtrack process. A new subclass of the CNF for which SAT is polynomial is evidentiated. Second, a new DPLL-like algorithm is proposed, based on unit resolution and a new dichotomous search. Finally, we devise a combinatorial approach based on renaming.

The rest of the paper is then devoted to the study of *weak transversals* by giving up the condition that the sets T^+ and T^- defining a transversal $T = (T^+, T-)$ of a directed hypergraph are disjoint. We are searching in this case for a weak transversal T minimizing $|T^+ \cap T^-|$. This is a very interesting problem giving rise to a new "degree of unsatisfiability" of a CNF formula, and having many decision combinatorial problems as instances. The section is ended with a structural result about unsatisfiable formulas. We conclude with a discussion section.

2   Directed Hypergraph Approach to SAT

It is not difficult to see that for SAT we can consider $\mathcal{C} = (C_i; i \in I)$ non empty, not containing the directed edge $(\emptyset, \emptyset)$, without directed edges C, C' satisfying $C \sqsubseteq C'$, and with the property that there is no partition (I_1, I_2) of I such that $\bigcup_{i \in I_1} V(C_i) \cap \bigcup_{i \in I_2} V(C_i) = \emptyset$. We call such a directed hypergraph $H = (V, \mathcal{C})$ a ***simple connected clutter***. If each edge $C \in \mathcal{C}$ satisfies $|V(C)| = k$, then H is a k-***uniform clutter***.

2.1   Clause Ordering

In order to capture the combinatorial properties of a simple connected clutter, we introduce the notion of *branching* that imposes an ordering of the members of the clutter, providing information about the associated transversal directed hypergraph.

Definition 1 *(C-Base Branching). Let $H = (V, \mathcal{C})$ be a simple connected clutter on V. A $\mathcal{C}$-**base branching** is any ordered set of edges of $\mathcal{C}$, $\mathcal{B} = \{B_1, \ldots, B_k\}$, such that:*

1. $V(B_i) - \bigcup_{j<i} V(B_j) \neq \emptyset$ for each $i = 1, \ldots, k$,
2. $V(B_i) \cap \bigcup_{j<i} V(B_j) \neq \emptyset$ for each $i = 2, \ldots, k$,
3. $\bigcup_{i=1,k} V(B_i) = V$.

$\mathcal{B}$ can be constructed by choosing B_1 and, at each step $i \in \{2, \ldots, k\}$, choosing a $\mathcal{C}$ member B_i, having in $V(B_i)$ at least a new element (cf. condition (1)) and at least a common element with $\bigcup_{j<i} V(B_j)$ (cf. condition (2)). The construction is possible since $\mathcal{C}$ is a simple connected clutter on V and finishes when $\bigcup_{i=1,k} V(B_i) = V$.

For each remaining edge $C \in \mathcal{C} - \mathcal{B}$ let $first(C, \mathcal{B})$ the first index $t \in \{1, \ldots, k\}$ such that $V(C) \subseteq \bigcup_{j=1,t} V(B_j)$. Clearly, $1 \leq first(C, \mathcal{B}) \leq k$ since, by the above definition of a base branching, we have $V(C) \subseteq V = \bigcup_{i=1,k} V(B_i)$.

It follows that each clause B_i in the base branching $\mathcal{B} = \{B_1, \ldots, B_k\}$ has associated a set of $\mathcal{C}$ members, $cov(B_i)$ from $\mathcal{C} - \mathcal{B}$, namely

$$cov(B_i) = (C \in \mathcal{C} - \mathcal{B} | first(C, \mathcal{B}) = i).$$

Clearly, $cov(B_i)$ and $cov(B_j)$ are disjoint for $i \neq j$, the union of all $cov(B_i)$ is $\mathcal{C} - \mathcal{B}$ and some $cov(B_i)$ can be empty. Note that $cov(B_i)$ designates the family of all members C out of the base branching with the property that $V(C) \subseteq \bigcup_{j=1,i} V(B_j)$ and $V(C) - \bigcup_{j=1,i-1} V(B_j) \neq \emptyset$.

Definition 2 *(Branching).* *If $H = (V, \mathcal{C})$ is a simple connected clutter on V, a **branching of** $\mathcal{C}$ is any ordering of the edges of $\mathcal{C}$: $B_1, cov(B_1), \ldots, B_k, cov(B_k)$, for a specified $\mathcal{C}$-base branching $\mathcal{B}$.*

Definition 3 *(C-Base Branching Depth).* *If $\mathcal{B} = \{B_1, \ldots, B_k\}$ is a $\mathcal{C}$-base branching, we define its **depth** by*

$$depth(\mathcal{B}) = \begin{cases} 0 & \text{if } \mathcal{C} - \mathcal{B} = \emptyset \\ \max\{i | cov(B_i) \neq \emptyset\} & \text{if } \mathcal{C} - \mathcal{B} \neq \emptyset. \end{cases}$$

Example. For H_F, the example of directed hypergraph given in the introduction, we can take $\mathcal{B} = \{B_1, B_2, B_3\}$, where $B_1 = C_1$, $B_2 = C_4$, $B_3 = C_6$, and therefore $cov(B_1) = \{C_2, C_3\}$, $cov(B_2) = \{C_4, C_5\}$, $cov(B_3) = \emptyset$. It follows that $depth(\mathcal{B}) = 2$.

Finding a $\mathcal{C}$-base branching with small depth is useful as the following proposition shows.

Proposition 1. *Let $H = (V, \mathcal{C})$ be a simple connected clutter on V and $\mathcal{B} = \{B_1, \ldots, B_k\}$ a $\mathcal{C}$-base branching. Then $Tr(H) \neq \emptyset$ if and only if*

$$Tr(\{B_1, cov(B_1), \ldots, B_{depth(\mathcal{B})}, cov(B_{depth(\mathcal{B})})\}) \neq \emptyset$$

Proof. We show how to transform a transversal

$$T \in Tr(\{B_1, cov(B_1), \ldots, B_{depth(\mathcal{B})}, cov(B_{depth(\mathcal{B})})\})$$

to a transversal

$$T' \in Tr(\{B_1, cov(B_1), \ldots, B_{depth(\mathcal{B})}, cov(B_{depth(\mathcal{B})}), B_{depth(\mathcal{B})+1}\}).$$

For $depth(\mathcal{B}) = 0$, the construction works by taking $T = 2\emptyset$. By the definition of the base branching, there is $v \in V(B_{depth(\mathcal{B})+1}) - \bigcup_{j \leq depth(\mathcal{B})} V(B_j)$. Now, T' can be constructed as

$$T' = \begin{cases} T \sqcup (\{v\}, \emptyset) & \text{if } v \in B^+_{depth(\mathcal{B})+1} \\ T \sqcup (\emptyset, \{v\}) & \text{if } v \in B^-_{depth(\mathcal{B})+1}. \end{cases}$$

A similar argument works for each $i \in \{depth(\mathcal{B}) + 1, \ldots, k - 1\}$ and thus we can extend T to a traversal of $\{B_1, cov(B_1), \ldots, B_{depth(\mathcal{B})}, cov(B_{depth(\mathcal{B})}), B_{depth(\mathcal{B})+1}, \ldots, B_k\} = \mathcal{C}$. $\qquad\square$

Note that if $depth(\mathcal{B}) = 0$, the above argument shows that $Tr(H) = Tr(\mathcal{C}) \neq \emptyset$. We have obtained the following:

Corollary 1. *If the edges of a directed hypergraph $H = (V, \mathcal{C})$ in a SAT instance can be ordered $\mathcal{C} = \{C_1, \ldots, C_m\}$ such that, for each $i \in \{2, \ldots, m\}$, $V(C_i)$ has at least a new element (that is, not belonging to $\bigcup_{j=1,i-1} V(C_j)$), then $Tr(\mathcal{C}) \neq \emptyset$.*

Moreover, it follows from the above proposition that, if $depth(\mathcal{B}) \leq c$ for some constant c (not depending on $|V|$ or $|\mathcal{C}|$), then we can test if $Tr(\mathcal{C}) \neq \emptyset$ by checking all partitions $P = (P^+, P^-)$ of $\bigcup_{j=1, depth(\mathcal{B})} V(B_j)$. If we find $P \in Tr(B_1, \ldots, B_{depth(\mathcal{B})})$, which intersects each member C in $\{cov(B_1), \ldots, cov(B_{depth(\mathcal{B})})\}$, then $Tr(\mathcal{C}) \neq \emptyset$. If each B_i has $|V(B_i)|$ constant (e.g. when $\mathcal{C}$ is 3-uniform) the number of these partitions is constant, and therefore the satisfiability of $\mathcal{C}$ can be tested in polynomial time:

Corollary 2. *Let $\mathcal{C}$ be a 3-uniform clutter of a SAT instance and $\mathcal{B} = \{C_1, \ldots, C_k\}$ a $\mathcal{C}$-base branching with $depth(\mathcal{B}) \leq c$, for some positive constant c. Then the satisfiability of $\mathcal{C}$ can be tested in polynomial time.*

The backtrack-free construction of a transversal suggested in the proof of Proposition 1, on which the Corollary 1 is based, can be explicitly described (in the general case) by the following algorithm:

---

1. Construct a $\mathcal{C}$-branching: $B_1, cov(B_1), \ldots, B_k, cov(B_k)$;
   $T \leftarrow B_1$;
   find $T' \in Tr(cov(B_1))$ such that $T \sqsubseteq T'$;
   $T \leftarrow T'$;
2. **for** $i = 2$ **to** k **do**
   Let $v \in V(B_i) - \bigcup_{j \leq i-1} V(B_j)$;
   $T \leftarrow \begin{cases} T \sqcup (\{v\}, \emptyset) & \text{if } v \in B^+_i \\ T \sqcup (\emptyset, \{v\}) & \text{if } v \in B^-_i. \end{cases}$
   find $T' \in Tr(cov(B_i))$ such that $T \sqsubseteq T'$;
   $T \leftarrow T'$;
3. **output** T.

The critical part of this algorithm is "find $T' \in Tr(cov(B_i))$ such that $T \sqsubseteq T'$" for each $i \in \{1, \ldots, k\}$. If $cov(B_i) = \emptyset$, as in the proof of Proposition 1 or in Corollary 1, we simply take $T' = T$. Another simple case is described in the following proposition.

Proposition 2. *Let $\mathcal{C}$ be a simple connected clutter on V and $\mathcal{B} = \{B_1, \ldots, B_k\}$ a $\mathcal{C}$-base branching. If for every $i \in \{1, \ldots, k\}$, each member $C \in cov(B_i)$ has an element $v(C, B_i) \in V(C) - \bigcup_{j=1, i-1} V(B_j)$ such that $v(C, B_i) \in C^+ \cap B_i^+ \cup C^- \cap B_i^-$, then $Tr(\mathcal{C}) \neq \emptyset$.*

Proof. By the definition of a $\mathcal{C}$-base branching, for every $i \in \{1, \ldots, k\}$ and $C \in cov(B_i)$, we have $V(C) - \bigcup_{j=1, i-1} V(B_j) \neq \emptyset$. The condition in hypothesis of the proposition assures that we can take T' in the above algorithm as

$$T' = T \bigsqcup_{\substack{C \in cov(B_i), \\ v(C, B_i) \in B_i^+}} (\{v(C, B_i)\}, \emptyset) \bigsqcup_{\substack{C \in cov(B_i), \\ v(C, B_i) \in B_i^-}} (\emptyset, \{v(C, B_i)\}).$$

We conclude this subsection by noting that the algorithm described can be easily integrated in a backtracking scheme using the branching base as a driver for the search. This could be useful for 3SAT since the extension of the current transversal T to one of $cov(B_i)$ can be done efficiently by solving a 2SAT.

2.2 A New DPLL-Like Algorithm for SAT

Let $H = (V, \mathcal{C})$ be a directed hypergraph and let $u, v \in V$, $u \neq v$. $H_{u=v} = (V - \{v\}, \mathcal{C}|_{u=v})$ is the directed hypergraph obtained from $H = (V, \mathcal{C})$ by the following algorithm:

```
C|u=v ← ∅;
for C ∈ C do {
        if v ∈ C⁺ then C⁺ ← C⁺ − {v} ∪ {u};
        if v ∈ C⁻ then C⁻ ← C⁻ − {v} ∪ {u};
        if u ∉ C⁺ ∩ C⁻ then C|u=v ← C|u=v ∪ {C};
        }
    return C|u=v.
```

In other words, $\mathcal{C}|_{u=v}$ is obtained from $\mathcal{C}$ by deleting all edges in which v and u appear on different sides, replacing v by u in all edges in which v appears but not u, and deleting v from the edges in which u and v appear on the same side.

Similarly, $H_{u \neq v} = (V - \{v\}, \mathcal{C}|_{u \neq v})$ is the hypergraph obtained from $H = (V, \mathcal{C})$ by the following algorithm:

$\mathcal{C}|_{u \neq v} \leftarrow \emptyset$;
for $C \in \mathcal{C}$ **do** {
    if $v \in C^+$ **then** { $C^+ \leftarrow C^+ - \{v\}$; $C^- \leftarrow C^- \cup \{u\}$ };
    if $v \in C^-$ **then** { $C^- \leftarrow C^- - \{v\}$; $C^+ \leftarrow C^+ \cup \{u\}$ };
    if $u \notin C^+ \cap C^-$ **then** $\mathcal{C}|_{u \neq v} \leftarrow \mathcal{C}|_{u \neq v} \cup \{C\}$;
    }
return $\mathcal{C}|_{u \neq v}$.

$\mathcal{C}|_{u \neq v}$ is obtained from $\mathcal{C}$ by deleting all edges in which v and u appear on the same side, replacing v by u and moving it to the other side in all edges in which v appears but not u, and deleting v from the edges in which u and v appear in different sides.

Clearly, $H_{u=v}$ and $H_{u \neq v}$ are (simplified) directed hypergraphs on $V - \{v\}$. The following proposition shows that this *elimination* of the vertex v is sound with respect to the existence of a transversal.

Proposition 3. $Tr(H) \neq \emptyset$ if and only if $Tr(H_{u=v}) \neq \emptyset$ or $Tr(H_{u \neq v}) \neq \emptyset$.

Proof. If $Tr(H) \neq \emptyset$, let $A = (A^+, A^-) \in Tr(H)$. We can suppose that $u, v \in A^+ \cup A^-$ (if $A = (A^+, A^-) \in Tr(H)$, then any pair A_1, with $A_1^+ \cap A_1^- = \emptyset$, $A^+ \subseteq A_1^+$ and $A^- \subseteq A_1^-$, is an edge of $Tr(H)$).

Case 1. $\{u, v\} \subseteq A^+$, or $\{u, v\} \subseteq A^-$. Suppose $u, v \in A^+$ and let $C_1 \in \mathcal{C}|_{u=v}$ such that $A^+ \cap C_1^+ \cup A^- \cap C_1^- = \emptyset$. It follows that $C_1 \notin \mathcal{C}$ and therefore C_1 is obtained from $C \in \mathcal{C}$, by replacing v by u in C^-. Since $u, v \in A^+$ it follows that $A^+ \cap C^+ \cup A^- \cap C^- = A^+ \cap C_1^+ \cup A^- \cap C_1^- = \emptyset$, contradiction. Hence $A \in Tr(H_{u=v})$. A similar argument can be used for the case $u, v \in A^-$.

Case 2. $u \in A^+, v \in A^-$ or $u \in A^-, v \in A^+$. Suppose $u \in A^+, v \in A^-$ and let $C_1 \in \mathcal{C}|_{u \neq v}$ such that $A^+ \cap C_1^+ \cup A^- \cap C_1^- = \emptyset$. It follows that $C_1 \notin \mathcal{C}$ and therefore C_1 is obtained from $C \in \mathcal{C}$, by deleting v from C^+ and adding u to C^-. Since $u \in A^+, v \in A^-$, it follows that $A^+ \cap C^+ \cup A^- \cap C^- = A^+ \cap C_1^+ \cup A^- \cap C_1^- = \emptyset$, contradicting the hypothesis that $A \in Tr(H)$. Therefore, $A \in Tr(H_{u \neq v})$. A similar argument can be used for the case $u \in A^-, v \in A^+$.

Conversely, if $Tr(H_{u=v}) \neq \emptyset$ or $Tr(H_{u \neq v}) \neq \emptyset$ then there is $A \in Tr(H_{u=v})$ or $B \in Tr(H_{u \neq v})$. We can suppose that $v \notin A^+ \cup A^-$ and $v \notin B^+ \cup B^-$, since no edge in $\mathcal{C}|_{u=v}$ or $\mathcal{C}|_{u \neq v}$ contains v. However, we can suppose that $u \in A^+ \cup A^-$ and $u \in B^+ \cup B^-$ (by adding it to A^+, respectively B^+, if necessary). We show that A and B can be transformed into transversals of H.

Transformation of A. If $u \in A^+$ then $A^+ \leftarrow A^+ \cup \{v\}$, else $A^- \leftarrow A^- \cup \{v\}$. Suppose that $u \in A^+$ and there is $C \in \mathcal{C}$ such that $A^+ \cap C^+ \cup A^- \cap C^- = \emptyset$. It follows that $C \notin \mathcal{C}|_{u=v}$, and this happens if $u \in C^-$ and $v \in C^+$, or $u \in C^+$ and $v \in C^-$. After the transformation of A, we have $u, v \in A^+$, and now $A^+ \cap C^+ \cup A^- \cap C^- \neq \emptyset$. A similar argument can be used for the case $u \in A^-$.

Transformation of B. If $u \in B^+$ then $B^- \leftarrow B^- \cup \{v\}$, else $B^+ \leftarrow B^+ \cup \{v\}$. Suppose that $u \in B^+$ and there is $C \in \mathcal{C}$ such that $B^+ \cap C^+ \cup B^- \cap C^- = \emptyset$. It follows that $C \notin \mathcal{C}|_{u \neq v}$, and this happens if $u, v \in C^-$ or $u, v \in C^+$. After the transformation of B, we have $v \in B^-$, and now $B^+ \cap C^+ \cup B^- \cap C^- \neq \emptyset$. Similarly, if $u \in B^-$. □

In order to use the above result in a backtracking method to find a transversal of a directed hypergraph, we need some notations. Let B and C pair of sets of V such that $B \sqcap C = 2\emptyset$. Then C^B denotes the pair C_1 with $C_1^+ = C^+ - B^-$ and $C_1^- = C^- - B^+$.

If $H = (V, \mathcal{C})$ is a directed hypergraph and B is a pair of sets of V such that $B^+ \cap B^- = \emptyset$, then

$$\mathcal{C}^B = (C^B | C \in \mathcal{C} \text{ and } (B^+ \cap C^+) \cup (B^- \cap C^-) = \emptyset)$$

Clearly, if $\mathcal{C}^B = \emptyset$ then $B \in Tr(H)$. Also, it is not difficult to see that there is $A \in Tr(H)$ extending B (i.e. $B^+ \subseteq A^+$ and $B^- \subseteq A^-$) if and only if $Tr(\mathcal{C}^B) \neq \emptyset$.

In particular, if $C \in \mathcal{C}$ is a unit clause then any transversal of $\mathcal{C}$ must extend C and, therefore, the following *unit propagation rule* holds:

$$Tr(\mathcal{C}) \neq \emptyset \text{ if and only if } Tr(\mathcal{C}^C) \neq \emptyset.$$

With our notations, the well known DPLL backtracking algorithm based on the unit propagation rule [2,3] can be modified to obtain a new complete algorithm for the SAT problem: VEB (*Variable Elimination Backtracking*).

---

procedure $VEB(\mathcal{C})$

(SAT)   **if** $\mathcal{C} = \emptyset$ **then return** *satisfiable*;

(Conflict)   **if** $(\emptyset, \emptyset) \in \mathcal{C}$ **then return** *unsatisfiable*;

(Unit Edge) **if** $\exists$ unit edge $C \in \mathcal{C}$ **then return** $VEB(\mathcal{C}^C)$;

(Branch) $(u, v) \leftarrow$ a pair of vertices of a minimum size edge of $\mathcal{C}$;

           if $VEB(\mathcal{C}|_{u=v})$ retur ns *satisfiable*

                    then return *satisfiable*

                    else return $VEB(\mathcal{C}|_{u \neq v})$

end.

2.3 Renaming

A very interesting approach to the SAT problem can be derived from the idea of *renaming* introduced in [13].

Let V be a finite nonempty set and $X \subseteq V$. For a pair $C = (C^+, C^-)$ of subsets of V, we define the X-*renaming of* C, denoted $r_X(C)$, the pair $r_X(C) = D = (D^+, D^-)$, where $D^+ = (C^+ - X) \cup (C^- \cap X)$ and $D^- = (C^- - X) \cup (C^+ \cap X)$ (that is, moving in C the elements of X from one side to another one).

Definition 4 *(Renaming).* *If $H = (V, \mathcal{C})$ is a directed hypergraph and $X \subseteq V$, then its X-**renaming** is the directed hypergraph $r_X(H) = (V, r_X(\mathcal{C}))$, where $r_X(\mathcal{C}) = (r_X(C)|C \in \mathcal{C})$.*

It is not difficult to see that $r_X(r_X(C)) = C$ and $C_1 \sqcap C_2 \neq 2\emptyset$ if and only if $r_X(C_1) \sqcap r_X(C_2) \neq 2\emptyset$. Also, $r_\emptyset(C) = C$, $r_V(C) = (C^-, C^+)$ and $r_X(r_Y(C)) = r_{X \triangle Y}(C)$, for every $X, Y \subseteq V$.

The following important property of renaming holds:

Proposition 4. *Let H be a directed hypergraph on V and $X \subseteq V$. Then $Tr(H) \neq \emptyset$ if and only if $Tr(r_X(H)) \neq \emptyset$. Moreover, $Tr(H) = r_X(Tr(r_X(H)))$.*

It follows that finding a transversal of H could be approached by finding a transversal of a suitable renaming $r_X(H)$ of it.

Definition 5 *(Transversal Renaming).* *A X-renaming of $H = (V, \mathcal{C})$ is called a **transversal** X-**renaming** if $r_X(C)^- \neq \emptyset$ for each $C \in \mathcal{C}$ or $r_X(C)^+ \neq \emptyset$ for each $C \in \mathcal{C}$.*

Proposition 5. *Let $H = (V, \mathcal{C})$ be a directed hypergraph. $Tr(H) \neq \emptyset$ if and only if H has a transversal X-renaming.*

Proof. If H has a transversal X-renaming then $T_1 = (\emptyset, V) \in Tr(r_X(H))$ or $T_2 = (V, \emptyset) \in Tr(r_X(H))$. By the above proposition, $r_X(T_1) = (X, V - X) \in Tr(H)$ or $r_X(T_2) = (V - X, X) \in Tr(H)$.

Conversely, if $T \in Tr(\mathcal{C})$ then it is easy to see that $X = T^+$ or $X = T^-$ is a transversal X-renaming of H. □

If $H = (V, \mathcal{C})$ is a directed hypergraph, let us denote $H_0^+ = (V, \mathcal{C}_0^+)$, where $\mathcal{C}_0^+ = (C \in \mathcal{C}|C^+ = \emptyset)$ and $H_0^- = (V, \mathcal{C}_0^-)$, where $\mathcal{C}_0^- = (C \in \mathcal{C}|C^- = \emptyset)$. We also denote $t(H) = \min(|\mathcal{C}_0^+|, |\mathcal{C}_0^-|)$.

Clearly, if $t(H) = 0$ then $Tr(H) \neq \emptyset$, since then $(V, \emptyset)$ or $(\emptyset, V)$ is a transversal of H. By the above proposition, it follows that

$$Tr(H) \neq \emptyset \text{ if and only if } \min_{X \subseteq V} t(r_X(H)) = 0.$$

Hence, we have thus reformulated the SAT problem on instance $H = (V, \mathcal{C})$, as the combinatorial search problem of a subset $X^* \subset V$ with the property that moving in each edge the vertices of X^* from one side to another, the number of edges having the same side nonempty is minimized. This opens the way to use in the SAT problem the well known heuristics methods of [5,11].

3   Weak Transversals

Let $H = (V, \mathcal{C} = (C_i; i \in I))$ be a directed hypergraph.

Definition 6 *(Weak Transversal).* *A **weak transversal** of $\mathcal{C}$ is a is a pair of subsets of V, $T = (T^+, T^-)$, such that $T \sqcap C_i \neq 2\emptyset$, for each $i \in I$.*

Clearly, every directed hypergraph H, not containing $2\emptyset$ as an edge, has a weak transversal. If $C, C' \in \mathcal{C}$ satisfies $C \sqsubseteq C'$, then every weak transversal of $H' = (V, \mathcal{C} - C')$ is a weak transversal of $H = (V, \mathcal{C})$. Therefore we can suppose that $H = (V, \mathcal{C})$ is a clutter (there is no $i, j \in I, i \neq j$ such that $C_i \sqsubseteq C_j$).

If T is a weak transversal of H and $T \sqsubseteq T^1$ then T^1 is also a weak transversal of H. A **minimal weak transversal** is a weak transversal T^0 of H such that if $T \sqsubseteq T^0$ and T is a weak transversal of H, then $T = T^0$. We denote by $Tr^w(H)$ the **clutter of minimal transversals** of H.

The following two simple lemmas are useful in order to prove that $Tr^w(Tr^w(H)) = H$, which could be interesting for some restrictions of the problem MinSAT bellow.

Lemma 1. *Let $H = (V, \mathcal{C})$ and $H' = (V, \mathcal{D})$ two clutters. If for each $C \in \mathcal{C}$ there is $D \in \mathcal{D}$ such that $C \sqsubseteq D$ and for each $D \in \mathcal{D}$ there is $C \in \mathcal{C}$ such that $C \sqsubseteq D$, then $H = H'$.*

Lemma 2. *Let $H = (V, \mathcal{C})$ be a clutter. If $X = (X^+, X^-)$ is a pair of subsets of V such that there is no $C \in \mathcal{C}$ with the property $C \sqsubseteq X$, then there is $T \in Tr^w(H)$ satisfying $X \cap T = 2\emptyset$.*

Proof. Let $\overline{X} = (V - X^+, V - X^-)$. Then $\overline{X}$ is a weak transversal of H. Otherwise, there is $C \in \mathcal{C}$ such that $C \sqcap \overline{X} = 2\emptyset$, that is, $C^+ \subseteq X^+$ and $C^- \subseteq X^-$, therefore $C \sqsubseteq X$, contradicting the hypothesis. It follows that there is a minimal weak transversal T of H such that $T \sqsubseteq \overline{X}$. Clearly, $X \cap T = 2\emptyset$. $\square$

We can now prove:

Proposition 6. *If $H = (V, \mathcal{C})$ is a clutter, then $Tr^w(Tr^w(H)) = H$.*

Proof. Let $C \in \mathcal{C}$. By the definition of $Tr^w(H)$ we have $C \sqcap T \neq 2\emptyset$ for every member T of $Tr^w(H)$. It follows that there is $D \in Tr^w(Tr^w(H))$ such that $D \sqsubseteq C$.

Let $X \in Tr^w(Tr^w(H))$. Then, there is $C \in \mathcal{C}$ such that $C \sqsubseteq X$. Otherwise, by Lemma 2, there is $T \in Tr^w(H)$ satisfying $X \cap T = 2\emptyset$, contradicting the assumption that $X \in Tr^w(Tr^w(H))$. The proposition follows now from Lemma 1. $\square$

Since a weak transversal T of directed hypergraph $H = (V, \mathcal{C})$ is a transversal of H if and only if $T^+ \cap T^- = \emptyset$, it follows that the following problem is NP-complete, containing (for $k = 0$) the problem SAT.

MinSAT

Instance: V a finite set, $\mathcal{C}$ a 2hypergraph on V, $k \in \mathbf{Z}_+$.

Question: Has $\mathcal{C}$ a weak transversal T such that $|T^+ \cap T^-| \leq k$?

It follows that we can take $\min\{|T^+ \cap T^-|; T \in Tr^w(H)\}$ as a measure of unsatisfiability of the directed hypergraph $H = (V, \mathcal{C})$. More precisely, we consider the following definition.

Definition 7 *(Unsatisfiability).* *Let* $H = (V, C)$ *a directed hypergraph. The* **unsatisfiability** *of* H, *denoted* $unsat(H)$, *is defined as:*

$$unsat(H) = \begin{cases} 0 & if\ C = \emptyset, \\ \min_{T \in Tr^w(H)} |T^+ \cap T^-| & if\ C \neq \emptyset\ and\ 2\emptyset \notin C, \\ \infty & if\ 2\emptyset \in C. \end{cases}$$

An obvious lower bound for $unsat(H)$ is the number of distinct pairs $(\{v_i\}, \emptyset), (\emptyset, \{v_i\})$ contained in C, because every weak transversal $T \in Tr^w(H)$ must contain $(\{v_i\}, \{v_i\})$. This observation can be easily extended to obtain the following proposition.

Proposition 7. *Let* $H = (V, C)$ *be a directed hypergraph, not containing* $2\emptyset$ *as an edge. If there is* $a \in V$ *such that* $(\{a\}, \emptyset), (\emptyset, \{a\}) \in C$, *then*

$$unsat(H) = 1 + unsat(H')$$

where $H' = (V, C')$ *and* $C' = \{C | C \in C, C \sqcap (\{a\}, \{a\}) = 2\emptyset\}$.

Some other properties of the unsatisfiability of a clutter are given in the next proposition.

Proposition 8. *Let* $H = (V, C)$ *be a clutter.*

1. *If* $H = (V, C')$ *is a clutter,* $H' = (V, C')$, *and* $C \subseteq C'$, *then* $unsat(H) \leq unsat(H')$.
2. *If* $H = Tr^w(H')$, *where* $H' = (V, C')$ *is a clutter, then* $unsat(H) = \min\{|C^+ \cap C^-|;\ C \in C'\}$.
3. *If* H *has an edge* $C \in C$ *which satisfies* $C^+ \cap C^- \neq \emptyset$, *then* $unsat(H) = unsat(H')$, *where* $H' = (V, C - C)$.
4. *If* $H' = (V', C')$ *is a clutter and* $V \cap V' = \emptyset$, *then* $unsat(H'') = unsat(H) + unsat(H')$, *where* $H'' = (V \cup V', C \cup C')$.

Proof.

1. Let $T \in Tr^w(H')$ such that $unsat(H') = |T^+ \cap T^-|$. Clearly T meets every member of C, therefore there is $T_1 \in Tr^w(C)$ such that $T_1 \sqsubseteq T$. It follows that $|T_1^+ \cap T_1^-| \leq |T^+ \cap T^-|$ and therefore $unsat(H) \leq unsat(H')$.
2. By Proposition 5, $Tr^w(H) = H'$ and the statement follows by the definition of $unsat(H)$.
3. By 1, we have $unsat(H') \leq unsat(H)$. Conversely, let $T \in Tr^w(H')$ such that $unsat(H') = |T^+ \cap T^-|$. If T meets C then, as above, we have $unsat(H) \leq unsat(H')$. If $T \sqcap C = 2\emptyset$ then, taking $v \in C^+ \cap C^-$, $T_1 = (T^+ \cup \{v\}, T^-)$ is a weak transversal of H. Any member $T_0 \in Tr^w(H)$, contained in T_1, satisfies $|T_0^+ \cap T_0^-| \leq |T_1^+ \cap T_1^-| \leq |T^+ \cap T^-|$, therefore $unsat(H) \leq unsat(H')$.
4. Since $V \cap V' = \emptyset$ it follows that $T \in Tr^w(H'')$ if and only if there are $T_1 \in Tr^w(H)$ and $T_2^w \in Tr(H')$ such that $T = T_1 \sqcup T_2$. The statement follows, since $|T^+ \cap T^-| = |T_1^+ \cap T_1^-| + |T_2^+ \cap T_2^-|$. □

The analogue of the resolution rule does not hold. Indeed, let $H = (V, \mathcal{C})$ and $C_1, C_2 \in \mathcal{C}$ such that there is $a \in C_1^+ \cap C_2^-$ and $C_1 \sqcup C_2 \neq (\{a\}, \{a\})$. If $R(C_1, C_2; a) = C_1 \sqcup C_2 \setminus (\{a\}, \{a\})$, then $unsat(H) \leq unsat(H')$, where $H' = (V, \mathcal{C} \cup R(C_1, C_2; a))$. This inequality follows from Proposition 8(1). Unfortunately, it could be strictly as the following example shows. Let $V = \{p, q, r\}$ and

$$\mathcal{C} = \{(\{p, q\}, \emptyset), (\{p, r\}, \emptyset), (\emptyset, \{p\}), (\emptyset, \{q\}), (\emptyset, \{r\})\}.$$

We then add to $\mathcal{C}$, by applying the above restricted resolution rule, the following members: $(\{p\}, \emptyset), (\{r\}, \emptyset), (\{q\}, \emptyset)$, obtaining $H' = (V, \mathcal{C}')$. Using Proposition 7, we obtain $unsat(H') = 3$, while $unsat(\mathcal{C}) = 1$, by taking the weak transversal $T = (\{p\}, \{p, q, r\})$.

The following proposition shows that, in order to compute $unsat(H)$ for a given directed hypergraph, we can transform it, in polynomial time, in a clutter with each edge having one side empty.

Proposition 9. *Let $H = (V, \mathcal{C})$ be a clutter and $C \in \mathcal{C}$ such that $|C^+|, |C^-| \geq 1$. If $v_C \notin V$, $V' = V \cup \{v_C\}$, $\mathcal{C}' = \mathcal{C} - C \cup (C^+ \cup \{v_C\}, \emptyset) \cup (\emptyset, C^- \cup \{v_C\})$, and $H' = (V', \mathcal{C}')$, then $unsat(H) = unsat(H')$.*

Proof. Let $T \in Tr^w(H)$ such that $unsat(H) = |T^+ \cap T^-|$. If T meets both C^+ and C^- then $T \in Tr^w(H')$. Otherwise, T meets exactly one of them and $T \sqcup (\emptyset, \{v_C\}) \in Tr^w(H')$ or $T \sqcup (\{v_C\}, \emptyset) \in Tr^w(H')$. Therefore $unsat(H) \leq unsat(H')$. Conversely, let $T \in Tr^w(H')$ such that $unsat(H') = |T^+ \cap T^-|$. If $v_C \notin T^+ \cap T^-$ then $T_1 = T - (\{v_C\}, \{v_C\}) \in Tr^w(H)$. Otherwise, $T^+ \cap C^+ = \emptyset$ and $T^- \cap C^- = \emptyset$. If $a \in V(C)$, then $T_1 = T - (\{v_C\}, \{v_C\}) \sqcup (\{a\}, \{a\}) \in Tr^w(H)$. It follows that $unsat(H') \leq unsat(H)$. □

As a consequence of the above proposition, we obtain that the MinSAT problem can be polynomially reduced to the following decision problem on (usual) hypergraphs:

Minimum intersection transversals
Instance: H_1 and H_2 two hypergraphs, $k \in \mathbf{Z}_+$.
Question: Are there T_i transversals of H_i ($i = 1, 2$) such that
$|T_1 \cap T_2| \leq k$?

Another interesting simplification of the directed hypergraph $H = (V, \mathcal{C})$, which does not change $unsat(H)$ is described in the following proposition (extending the corresponding polynomial reduction of SAT given in [14]). Essentially, it states that we can suppose that any two members of $\mathcal{C}$ have at most one element in their corresponding support sets. We use the following notation: if X is a pair of subsets of V, $a \in V$ and $b \notin V$, then the pair on $V \cup \{b\}$ obtained from X by replacing a by b is denoted $X[a \leftarrow b]$.

Proposition 10. *Let $H = (V, \mathcal{C})$ be a directed hypergraph and $C_1, C_2 \in \mathcal{C}$ such that there is $a \in V(C_1) \cap V(C_2)$. If $b, c \notin V$, $b \neq c$, $V_1 = V \cup \{b, c\}$, $\mathcal{C}_1 = \mathcal{C} - \{C_1, C_2\} \cup \{C_1[a \leftarrow b], C_2[a \leftarrow c]\} \cup \{(\{a\}, \{b\}), (\{b\}, \{c\}), (\{c\}, \{a\})\}$ and $H_1 = (V_1, \mathcal{C}_1)$, then $unsat(H) = unsat(H_1)$.*

Proof. Let $T \in Tr^w(H)$ such that $unsat(H) = |T^+ \cap T^-|$. We extend T to T_1, a pair of subsets of V_1, which meets each edge of $\mathcal{C}_1$. If $a \in T^+$ then $T_1 = T \cup (\{b\}, \emptyset) \cup (\{c\}, \emptyset)$. If $a \in T^-$ then $T_1 = T \cup (\emptyset, \{b\}) \cup (\emptyset, \{c\})$. Finally, if $a \notin V(T)$ then $T_1 = T \cup (\{a\}, \emptyset) \cup (\{b\}, \emptyset) \cup (\{c\}, \emptyset)$. In all cases we have $|T_1^+ \cap T_1^-| = |T^+ \cap T^-|$, therefore it follows that $unsat(H_1) \leq unsat(H)$.
Conversely, let $T \in Tr(H_1)$ such that $unsat(H_1) = |T^+ \cap T^-|$. If we take $T' = T[b \leftarrow a]$ and $T_1 = T'[c \leftarrow a]$, it is not difficult to see that T_1 meets each member of $\mathcal{C}$. Moreover, the substitutions $T[b \leftarrow a]$ and $T'[c \leftarrow a]$ do not increase $|T_1^+ \cap T_1^-|$, because either $a \in T^+ \cap T^-$ or b, c are forced to be in $T^+ - T^-$ or in $T^- - T^+$, by the structure of the $\mathcal{C}_1$. It follows that $unsat(H) \leq unsat(H_1)$. □
    A known measure of unsatisfiability of a directed hypergraph $H = (V, \mathcal{C})$ is to consider (via MAXSAT problem) $Unsat(H)$, the minimum number of edges which must be deleted from $\mathcal{C}$ such that the resulted directed hypergraph has a transversal:

$$Unsat(H) = |\mathcal{C}| - \max\{|\mathcal{C}'|; \mathcal{C}' \subseteq \mathcal{C} \text{ and } Tr(\mathcal{C}') \neq \emptyset\}.$$

    In the logical formulation of SAT, $Unsat(H)$ is the minimum number of unsatisfied clauses from $\mathcal{C}$, over all possible truth assignments of the variables in V.
    Our parameter $unsat(H)$ refers to the minimum number of "trouble" variables which must be "considered" both *true* and *false* in order to satisfy all clauses. More precisely, let V' be a disjoint copy of V. For each $v \in V$ let $\mathcal{C}_v^+$ $(\mathcal{C}_v^-)$ be the set of all clauses containing v (respectively, $\neg v$). If in each clause $C \in \mathcal{C}_v^-$ we substitute $\neg v$ by the copy v' of v, obtaining C', then all clauses in $\mathcal{C}_v^+ \cup (\mathcal{C}_v^-)'$ can be made true by setting v and v' to *true*. It is not difficult to see that the minimum number of copies of the variables in V, which must be considered in order to satisfy all clauses in $\mathcal{C}$, is exactly $unsat(H)$.
    It is well-known [9] that determining $Unsat(\mathcal{C})$ is a NP-hard problem even each clause in $\mathcal{C}$ has exactly 2 literals. A similar conclusion can be derived for $unsat(\mathcal{C})$ from the following proposition, using VERTEX COVER [8].

Proposition 11. *Let $G = (V, E)$ be a graph and $\tau(G)$ its vertex covering number. If $H_G = (V, \mathcal{C}_G)$ is the directed hypergraph with $\mathcal{C}_G = ((\{v\}, \emptyset))_{v \in V} \cup ((\emptyset, \{v, w\})_{vw \in E})$, then $\tau(G) = unsat(H_G)$.*

Proof. Let $T \in Tr^w(H_G)$. Since T meets each $(\{v\}, \emptyset))_{v \in V}$ member of $\mathcal{C}_G$, it follows that $T^+ = V$. Since T meets each $(\emptyset, \{v, w\})_{vw \in E}$ member of $\mathcal{C}_G$, it follows that T^- is a vertex cover of G. Therefore $unsat(H_G) = \min_{T \in Tr^w(\mathcal{C}_G)} |T^+ \cap T^-| = \min_{T \in Tr^w(\mathcal{C}_G)} |T^-| = \min_{T^- \text{ vertex cover of } G} |T^-| = \tau(G)$. □
    In the next proposition, the two unsatisfiability degrees are compared, proving that our newly introduced degree is not greater than the usual one.

Proposition 12. *Let $H = (V, \mathcal{C})$ be a directed hypergraph not containing $2\emptyset$. Then $unsat(H) \leq Unsat(H)$.*

Proof. Let $Unsat(H) = k$, $\mathcal{C} = \mathcal{C}' \cup \{C_1, \ldots, C_k\}$, $H' = (V, \mathcal{C}')$, and $T \in Tr(H')$. By the definition of k, it follows that for each $i \in \{1, \ldots, k\}$, every $v \in V(C_i)$ has been used in T and either $v \in C_i^+ \cap T^-$ or $v \in C_i^- \cap T^+$. Hence, choosing from each $V(C_i)$ an element v_i, we have a weak transversal of H, $T_0 = T \bigsqcup_{i=1,k}(\{v_i\}, \{v_i\})$, with the property that $|T_0^+ \cap T_0^-| = k$. Hence $unsat(H) \leq k$ and the proposition holds. $\square$

The above inequality can be used to determine $Unsat(H)$ for particular directed hypergraphs or to establish combinatorial relations. Some examples are given bellow:

1. Let $H_G = (V, \mathcal{C}_G)$ be the directed hypergraph associated to the graph $G = (V, E)$ in Proposition 10. $unsat(H_G)$ is the vertex covering number $\tau(G)$. In order to determine $Unsat(H_G)$, observe that taking T^- a minimum cardinality vertex cover of G and $T^+ = V - T^-$ we obtain a transversal T of $H = (V, \mathcal{C}_G - ((\{v\}, \emptyset))_{v \in T^-})$, therefore $Unsat(H_G) \leq \tau(G)$. Hence, by Proposition 12, we have $unsat(H_G) = Unsat(H_G) = \tau(G)$.

2. Let $G = (R, S; E)$ be a bipartite graph and let $H_G = (V, \mathcal{B}_G)$ be the following directed hypergraph on $V = R \cup S$: $\mathcal{B}_G = ((\{r\}, \emptyset))_{r \in R} \cup (\emptyset, \{s\}))_{s \in S} \cup ((\{s\}, \{r\}))_{rs \in E}$. Let $T \in Tr^w(\mathcal{B}_G)$. Clearly, $R \subseteq T^+$ and $S \subseteq T^-$. Therefore $T^+ = R \cup S_0$, $T^- = S \cup R_0$, where $R_0 \subseteq R$, $S_0 \subseteq S$ and $R_0 \cup S_0$ is a vertex cover of G. It follows that $unsat(H_G) = \tau(G)$. If $M \subseteq E$ is a maximum matching in G let $R_M = \{r \in R | \exists s \in S \text{ such that } rs \in M\}$ and $S_M = \{s \in S | \exists r \in R \text{ such that } rs \in M\}$. It is easy to see that $T_M = (R_M, S_M)$ satisfies $T_M \in Tr(H)$, for $H = (R \cup S, \mathcal{B}_G - ((\{r\}, \emptyset))_{r \in R - R_M} \cup (\emptyset, \{s\}))_{s \in S - S_M})$. Furthermore, since $R - R_M \cup S - S_M$ is a vertex cover of G, we have $Unsat(H_G) \leq \tau(G)$. Hence, by Proposition 12, we have $unsat(H_G) = Unsat(H_G) = \tau(G)$.

3. Let $G = (V, E)$ be a graph and let $H_G = (V, \mathcal{D}_G)$ be the following 2 directed hypergraph: $\mathcal{D}_G = ((\{v, w\}, \emptyset))_{vw \in E} \cup ((\emptyset, \{v, w\}))_{vw \in E}$. It is easy to see that $Tr(\mathcal{D}_G) \neq \emptyset$ if and only if G is a bipartite graph. If $T \in Tr^w(\mathcal{D}_G)$, then T^+ and T^- are vertex covers of G, therefore

$$unsat(H_G) = n - \max\{|S_1 \cup S_2| \; ; \; S_1, S_2 \text{ stable sets in } G\}.$$

On the other hand, it is not difficult to see that $Unsat(H_G) = m - maxcut(G)$. Using Proposition 12, we obtain an interesting inequality which holds in the given graph.

The difference between the two unsatisfiability degrees can be made no matter how big, as the following proposition states.

Proposition 13. *For each positive integer n there is a directed hypergraph $H = (V, \mathcal{C})$ with $|V| = 2n$ vertices and $|\mathcal{C}| = 8n$ edges such that $Unsat(H) = n + unsat(H)$.*

Proof. Let $H_1 = (\{v_1, v_2\}, \mathcal{C})$, where
$\mathcal{C}_1 = \{(\emptyset, \{v_1\}), (\emptyset, \{v_2\}), (\emptyset, \{v_1, v_2\}), (\{v_1\}, \emptyset), (\{v_2\}, \emptyset), (\{v_1, v_2\}, \emptyset), (\{v_1\}, \{v_2\}), (\{v_2\}, \{v_1\})\}$. $T = (\{v_1, v_2\}, \{v_1, v_2\})$ is a weak transversal of H_1 with $|T^+ \cap T^-| = 2 = unsat(H_1)$. However, it is not difficult to see (by inspection) that $Unsat(\mathcal{C}_1) = 3$.

Let H be the union of n disjoint copies of the directed hypergraph H_1. The proof follows from Proposition 8(4).   □

The next proposition shows that any non-trivial directed hypergraph H with no tansversal is the disjoint union of a directed hypergraph having transversals and a directed hypergraph with exactly $unsat(H)$ vertices such that any transversal of the first one and any vertex of the second one satisfy an obvious conflicting property, preventing the (partial) transversal to be extended to a transversal of H.

Proposition 14. *Let $H = (V, \mathcal{C})$ be a directed hypergraph not containing $2\emptyset$. $Tr(H) = \emptyset$ if and only if there is $X \subset V$, $X \neq \emptyset$, such that, if $H - X = (V, (C \in \mathcal{C}|V(C) \cap X = \emptyset))$, then:*

– *$Tr(H - X) \neq \emptyset$ and*
– *for every $T \in Tr(H - X)$ and every $x \in X$ there are $C_1, C_2 \in \mathcal{C}$ such that $x \in C_1^+ \cap C_2^-$, $C_1^+ - \{x\}, C_2^+ \subseteq T^-$ and $C_1^-, C_2^- - \{x\} \subseteq T^+$.*

Proof. Suppose that the condition in the proposition holds and there is $T \in Tr(H)$. Clearly, $T \in Tr(H - X)$. Let $x \in X$. It follows that there are $C_1, C_2 \in \mathcal{C}$ such that $x \in C_1^+ \cap C_2^-$, $C_1^+ - \{x\}, C_2^+ \subseteq T^-$ and $C_1^-, C_2^- - \{x\} \subseteq T^+$. But then, either $C_1 \sqcap T = 2\emptyset$ or $C_2 \sqcap T = 2\emptyset$, contradicting $T \in Tr(H)$.
Conversely, suppose $Tr(H) = \emptyset$. Since $2\emptyset \notin \mathcal{C}$, it follows that $0 < unsat(H) < \infty$. Let $T_0 \in Tr^w(H)$ with $|T_0^+ \cap T_0^-| = unsat(H)$ and let us consider $X = T_0^+ \cap T_0^-$. Clearly, $X \neq \emptyset$. Furthermore, $\hat{T} = T - (X, X)$ is a transversal of $H - X$. Hence $Tr(H - X) \neq \emptyset$. Also, for every $T \in Tr(H - X)$, $T_1 = T \sqcup (X, X)$ is a weak transversal of H satisfying $|T_1^+ \cap T_1^-| = |X| = unsat(H)$.
Let $H_1 = (X, (C - (X, X)|C \in \mathcal{C}, C \sqcap T = 2\emptyset))$. Then, for each $x \in X$ both $(\{x\}, \emptyset)$ and $(\emptyset, (\{x\})$ are edges of H_1. Indeed, if $(\{x\}, \emptyset)$ is not an edge of H_1, for some $x \in X$, then $T_2 = T_1 - (\{x\}, \emptyset)$ satisfies $T_2 \in Tr^w(H)$ and $|T_2^+ \cap T_2^-| < |T_1^+ \cap T_1^-| = unsat(H)$, a contradiction. A similar argument shows that $(\emptyset, (\{x\})$ is an edges of H_1.
The condition in the statement of the theorem holds, since $(\{x\}, \emptyset)$ and $(\emptyset, (\{x\})$ are edges of H_1 if and only if there are $C_1, C_2 \in \mathcal{C}$ such that $x \in C_1^+ \cap C_2^-$, $C_1^+ - \{x\}, C_2^+ \subseteq T^-$ and $C_1^-, C_2^- - \{x\} \subseteq T^+$.   □

4 Discussion

Existing literature reports several hypergraph formulation of the SAT problem [4,10,12,14]. In the directed hypergraph setting [6,7], the SAT problem has been formulated as the problem of finding a hyperpath between two specified nodes, while the problem of finding the minimum number of clauses to be deleted in order to make a formula satisfiable was formulated as the problem of finding a minimum cutset.

Our approach is focused on the connection between SAT problem and transversals in directed hypergraphs, being fairly direct and simple, and having the advantage of an uniform addressing of clauses and assignments.

The present paper has highlighted new possibilities of addressing the SAT problem within this framework. We have shown that, exploiting the directed hypergraph representation induced combinatorial properties and techniques, new problems, ideas and results arise. Clause ordering helped highlighting a new polynomial class of formulae for which SAT is polynomial. The vertices elimination algorithm introduced for finding transversal in a directed hypergraph put forward a natural way of variable elimination not considered (as we know) in the SAT literature. An interesting translation of the SAT problem into a combinatorial search problem, which could be approached by usual heuristic of the field, is also described (via renaming). Finally the relaxation approach taken by the weak transversals (intuitively obvious in this alternative syntactic setting) gave interesting measures of unsatisfiability directly connected to decision problems on graphs (e.g., vertex cover).

Despite the (sometimes) cumbersome notations, this paper has a rigorous contribution and message. It opens the path to new approaches to the SAT problem that are "easy" to obtain using the directed hypergraph representation and not intuitive otherwise. The above main results obtained are encouraging in this sense. They allow opti

References

1. Ausiello, G.: Directed hypergraphs: data structures and applications. In: Dauchet, M., Nivat, M. (eds.) CAAP 1988. LNCS, vol. 299, pp. 295–303. Springer, Heidelberg (1988)
2. Davis, M., Logemann, G., Loveland, D.: A machine program for theorem-proving. Commun. ACM **5**(7), 394–397 (1962)
3. Davis, M., Putnam, H.: A computing procedure for quantification theory. J. ACM **7**, 201–215 (1960)
4. Eiter, T., Gottlob, G., Makino, K.: New results on monotone dualization and generating hypergraph transversals. In: Proceedings of 34th ACM Symposium on Theory of Computing, Montreal, Quebec, Canada, 19–21 May 2002
5. Fiduccia, C.M., Mattheyses, R.M.: A linear time heuristic for improving network partitions. In: Proceedings of ACM/IEEE Design Automation Conference, pp. 175–181 (1982)
6. Gallo, G., Gentile, C., Pretolani, D., Rago, G.: Max Horn sat and the minimum cut problem in directed hypergraphs. Math. Program. **80**, 213–237 (1998)
7. Gallo, G., Longo, G., Pallottino, S., Nguyen, S.: Directed hypergraphs and applications. Discrete Appl. Math. **42**, 177–201 (1993)
8. Garey, M., Johnson, D.: Computers and Intractability: A Guide to the Theory of NP-Completeness. Freeman, New York (1979)
9. Garey, M., Johnson, D., Stockmeyer, L.: Some simplified np-complete graph problems. Theor. Comput. Sci. **1**, 237–267 (1976)

10. Kavvadias, D., Papadimitriou, C.H., Sideri, M.: On Horn envelopes and hyper-graph transversals. In: Ng, K.W., Raghavan, P., Balasubramanian, N.V., Chin, F.Y.L. (eds.) Algorithms and Computation. LNCS, vol. 762, pp. 399–405. Springer, Heidelberg (1993)
11. Kernighan, B.W., Lin, S.: An efficient heuristic procedure for partitioning graphs. Bell Syst. Tech. J. **49**, 291–307 (1970)
12. Kullmann, O.: An application of matroid theory to the sat problem. Technical report, ECCC TR00-018 (2000)
13. Lewis, H.: Renaming a set of clauses as a horn set. J. Assoc. Comput. Mach. **25**, 134–135 (1978)
14. Porschen, S., Speckenmeyer, E., Randerath, B.: On linear CNF formulas. In: Biere, A., Gomes, C.P. (eds.) SAT 2006. LNCS, vol. 4121, pp. 212–225. Springer, Heidelberg (2006)
15. Torres, A.F., Araoz, J.D.: Combinatorial models for searching in knowledge bases. Mathematicas Acta Cient. Venez. **39**, 387–394 (1988)

Conceptual Graphs for Formally Managing and Discovering Complementary Competences

Nacer Boudjlida[1]([✉]) and Badrina Guessoum-Boumezoued[2]

[1] Loria, Lorraine University, Nancy, France
Nacer.Boudjlida@loria.fr
[2] Bejaia University, Bejaia, Algeria
badrinagasmi@gmail.com

Abstract. The capture, the structuring of the expertise or the competences of an "object" (lie a business partner, an employee and even a software component or a Web service) are of very crucial interest in many application domains, like cooperative and distributed applications as well as in cooperative e-business applications and in human resource managment. The work that is described in this paper concerns the advertising, the classification and the discovry of competences. The foundings of the proposals that are described here after are a formal representation of competences using conceptual graphs and the use of operations on conceptual graphs for competence discovery and their possible composition.

Keywords: Competence management · Conceptual graphs · Competence discovery · Competence complementarity

1 Introduction

A competence management process [1] can be achieved following three steps: (1) *Competence identification*: it consists in describing competences under a formal representation. (2) *Competence organization*: once represented, competences are organized, classified and structured in order to be efficiently exploited and (3) *Competence use*: it consists in exploiting the organized competences. In this work, we aim at exploiting the competences for their discovery, i.e. when searching for entities that meet given needs.

Competence management and discovery find their application in different domains, like component-based programming, semantic-based Web services discovery [27], e-business, human resources management and even enterprise knowledge management [17]. For example, in the e-business domain, we see the application of our work when seeking for possible partners or subcontractors. In human resource management, considering employees enrollment as an example, the application of our wok can be useful when looking for employees satisfying a given work position profile.

In this paper, we aim at proposing a generic approach which can be instantiated in different domains. The ultimate goal is to define a method for competence

© Springer International Publishing Switzerland 2015
M. Croitoru et al. (Eds.): GKR 2015, LNAI 9501, pp. 89–106, 2015.
DOI: 10.1007/978-3-319-28702-7_6

management and apply the method for competence discovery and composition in distributed knowledge bases. A significant originality of the proposed approach resides in the type of answers we aim at providing. Indeed, when no unique entity satisfies the search criteria, the system attempts to determine a composite answer, i.e. a set of entities that satisfy the whole search criteria, every entity in the resulting set satisfying part of the criteria.

For competence representation and management, we rely on a knowledge representation using Conceptual Graphs (CGs) [14]: we not only represent knowledge as graphs but the reasoning is made thanks to graph-based operations. From a system architecture point of view, we use a mediator-based architecture [5], i.e. a set of distributed and cooperative mediators.

The presentation of this work is structured as follows. Section 2 presents related work and the work background. Section 3 presents the proposed approach for competence management and discovery. Section 4 provides an overview of the implementation of the approach whereas concluding remarks are in Sect. 5.

2 Related Work and Background

The current work is related to three main bodies of research: (i) Knowledge Representation (Sect. 2.1), (ii) competence representation and discovery (Sects. 2.2 and 2.3) and (iii) heterogeneous and distributed architectures (Sect. 2.4). We briefly discuss important studies in these research areas.

2.1 Knowledge Representation

During the past 40 years, a wide variety of Knowledge Representation (KR) formalisms has been developed. In general, these formalisms fall into two categories: (1) those that follow a "logical approach" (like Description Logic [25]) and provide a general reasoning machinery and a representation language which is usually a variant of the first-order predicate calculus and (2) those that follow a "non-logical approach" (like Semantic Networks [20] and CGs [8,14]) that use graphical interfaces that enable representing knowledge manipulation according to *ad-hoc* data structures. CGs are briefly introduced hereafter.

CGs are presented as a general model for knowledge representation. They were conceived to represent the semantics of natural languages; they evolved to become complete systems in the sense of logic. A CG description represents **ontological knowledge** in a structure called *support* which introduces the vocabulary of the studied domain. The *support* is implicitly used in the representation of **factual knowledge** as labeled graphs called conceptual graphs.

The support consists of (an example is in Fig. 1) *(i) a hierarchy of concept types* organized around the relation of specialization/generalization, *(ii) a set of relation types* organized into several hierarchies, each of them organizes relation types having the same arity, *(iii) a set of markers or referents* (denoted by I in Fig. 1) that refers to specific concepts (an unspecified concept can be referenced using a generic marker denoted as *), *(iv) a conformity relation* (τ in Fig. 1)

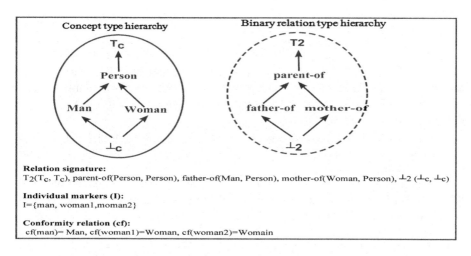

Fig. 1. Conceptual graph support

which relates markers to concept types and *(v) signatures of relations* which represent all the graphs which express constraints associated with every relation. A signature defines the number of the relation's arguments and their types. A graph signature is constituted by elementary graphs from which we can construct more complex graphs.

Furthermore, a CG is composed of: (1) A set of concept-nodes labeled from a support. A concept is composed of a *referent* that identifies the represented object, a type which classifies the represented object and (2) a set of relation-nodes labeled from a support. A relation is composed of a label which identifies the type of the relation and a set of edges linking the relation to its related concepts.

CGs can have different concrete notations such as graphical representation, textual notation and Conceptual Graph Interchange Format (CGIF) [24].

In a graphical notation, called display form (DF) (see Fig. 2), concepts are represented by rectangles and relations are represented by circles or ovals. The arcs that link the relations to the concepts are represented by arrows.

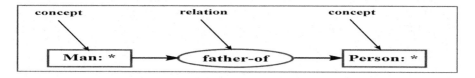

Fig. 2. A conceptual graph example

In a textual notation, called linear form (LF), concepts are represented by square brackets and relations are represented by parenthesis. Under a LF notation, the CG of Fig. 2 is expressed as: [Man: *] → (father-of) → [Person: *].

The CGIF notation has a syntax that uses co-reference labels to represent the arcs. The example in Fig. 2 is expressed in CGIF as: [Man: *m] [Person: *p] (father-of ?m ?p). *m and *p are variable definitions and ?m and ?p are references to defined variables.

A CGs being a logic system, it can easily be translated under a predicate logic form. As an example, the CG in Fig. 2 is expressed as: $\exists$ m $\exists$ p: Person(p) $\wedge$ Man(m) $\wedge$ father-of(m,p).

Furthermore, a variety of operations and extensions [8] are defined on CGs. We recall hereafter those that are necessary to the comprehension of the remainder of this paper.

- **Projection** is defined as an application $\prod$ of the nodes of a graph H towards the nodes of a graph G such as: (1) for each concept c in H, $\prod$(c) is either a specialization or the same as c, (2) for each relation r in H, $\prod$(r) is either a specialization or the same as r, (3) if the ith edge of r is linked to a concept c in H, then the ith edge of $\prod$(r) must be linked to $\prod$(c) in G.
- The **Normalization operation** returns a graph under a normal form which respects a structure where the markers are unique by merging concepts having the same individual marker. The normal form of a graph avoids semantic and logical ambiguity in CGs. Formally, let H be a CG, and C be the set of its concepts. H is under its normal form if for each couple of concepts (c1, c2) c1 and c2 $\in$ C, referent(c1) $\neq$ referent(c2).
- The **Disjoint sum** consists in drawing another CG next to the original CG [16]. Formally, let H1 and H2 be two CGs, and let (C1, R1, E1) and (C2, R2, E2) the concept set, the relation set and the edge set of H1 and H2 respectively. The disjoint sum of H1 and H2 is a CG H(C, R, E) such as (1) C is the union of C1 and C2, (2) R is the union of R1 and R2 and (3) E is the union of E1 and E2.
- **Headed graphs** are graphs that have a certain node chosen as the semantic head.
- **Conceptual graph rules** [22] were proposed as an extension of simple CGs to represent "IF A THEN B" knowledge where A and B are simple CGs. Formally, a graph rule is constituted from an hypothesis graph A, a conclusion graph B and a set of attach points corresponding to connection links between A and B. The rule application mechanism in a CG is based on the projection operation.

2.2   Competence Representation

Competence representation is a sub-field of KR which extends current KR languages to be more suited for competence description [9]. In [3], competences are methods of object-oriented software. Furthermore, DL is used to describe the intended semantics of these objects and the possible constraints involving their methods. In [4], entities are software objects and competences are the capabilities of a software object. In [5], entities are a set of activities (or functions) describing a given domain, an activity being described by the set of the required competences to carry it out. These competences represent the set of properties (or attributes) of the activities and their intended semantic is expressed using DL.

2.3   Competence Discovery

Competence discovery consists in searching entities having a set of required competences in order to satisfy a given objective. Answers to a competence discovery request may be of two types: (1) single answers, when single entities satisfy the search criteria, (2) cooperative or composite answers when no single entity, but a set of entities, meets the search criteria. In [9], competence discovery is defined as a query-answer process that attempts to find out which kind of entities owns a competence, and who they are. In [5], a request X is viewed in term of DL language as a concept having the given competences and the request evaluation consists in locating this concept in the concept classification hierarchy. The answers of a request are the individuals or the instances of all the concepts subsuming X. In an extended work [6], the authors present a method to produce composite answers thanks to the notion of "complementary objects" that is founded on the complement concept in DLs [23].

2.4   Heterogeneous and Distributed Architectures

In order to satisfy a competence search request in an heterogeneous and distributed environment like Internet, we have to cope with competence descriptions expressed in different formalisms either locally or remotely. This facility requires techniques to transform a competence description from one formalism into another, together with communication between the systems managing the various competence descriptions. Different heterogeneous and distributed architectures are candidate to the implementation of these systems, like Service Oriented Architectures, Peer to Peer (P2P) architectures [2,15] and Mediator-based architectures, the latter being the one we rely on.

   A mediation architecture [29] tries to solve the problem of the access and the integration of information by introducing the notion of a mediator as "a software module that exploits encoded knowledge about some sets or subsets of data to create information for a higher layer of application". The mediation can be of two types:

- **Centralized mediation:** where only one mediator is considered. In this case, all the data sources are stored in the same base.
- **Distributed mediation:** (or federation of mediators) in which a set of mediators agree to be considered as a single entity when applications demand for services to the federation. Distributed mediation systems have become a reference architecture to integrate both structured and semi-structured data [18,29].

In addition, many mediator-based approaches have been proposed in the literature. In [28], a single mediator is designed to offer an adequate level of decision-making integration of heterogeneous computer systems. The Conflict Resolution Environment for Autonomous Mediation (CREAM) system has been implemented and it provides various user groups with an integrated and collaborative facility to achieve semantic interoperability among participating heterogeneous information sources [21]. The KRAFT (Knowledge Reuse And Fusion/Transformation) architecture provides a generic infrastructure for knowledge management applications. It supports virtual organization using mediator agents [19]. In [5,6,10,11], an architecture based on a heterogeneous federation of mediators has been adopted. In this architecture, great emphasis is on cooperation and heterogeneity aspects.

Now, let us turn toward our actual proposal for competence management and discovery.

3   Proposal: Competence Management and Discovery Using CGs

In this section, we firstly present how a request is reprented as a CG and, secondly, the request's satisfaction process. The mediator-based architecture, as well as the system architecture, will be described in Sect. 4.

3.1   Conceptual Architecture

In the proposed approach, a mediator-based architecture has been adopted as described in [5]. It is very similar to the notion of discovery agency in the Web service architecture [11]. In this architecture, an "entity", called exporter, publishes its competences at one or more mediators (arrow (a) in Fig. 3). Entities, called importers, send requests to the mediator asking for exporters fitted with a given set of competences (arrow (b) in Fig. 3). The mediator explores its competence base to try to satisfy the request. The competence search process is founded on the exported competences and on relationships between them, these relationships being transparently established by the mediator. When the request can be satisfied by some exporters, the references of these exporters are sent back to the importer (arrow (c) in Fig. 3).

In this architecture, some cases may conduct to a failure of the request when only one mediator is involved. But, if we assume a grouping of mediators, these cases are typical cases where cooperation of mediators is required. When a mediator partner fails in the satisfaction of a request, we need to determine what is missing to the entities to satisfy request. That missing part is then transmitted to a mediator in the federation who, in turn, behaves like the preceding mediator. Therefore, satisfying a request may fall under different cases [9]:

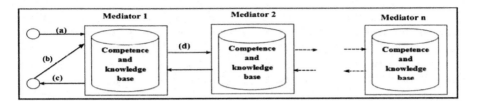

Fig. 3. The mediator-based architecture

1. there exist exporters that fully satisfy the request;
2. there exist exporters that partly satisfy the request but, when "combining" or composing the competences of different exporters one can fully satisfy the request;
3. no single exporter nor multiple exporters satisfy the request. In the latter situation, the mediator may initiates a cooperation process with other mediators to attempt to satisfy the request (arrow (d) in Fig. 3).

In addition, in a federated mediator architecture, the competence discovery can fall under the following situations:

1. *Homogeneous local satisfaction* where the request and the knowledge base are in the same KR language, and the knowledge base is located in one server.
2. *Homogeneous distributed satisfaction:* where the request and the knowledge base are in the same KR language, and the knowledge base is distributed in several servers.
3. *Heterogeneous satisfaction:* where the request and the knowledge base are in different KR languages and the knowledge base may be distributed.

In this work, we only deal with the homogeneous distributed satisfaction.

3.2 Competence Representation

Using CGs, competences are represented by relations and entities are represented by concepts. For example, saying that a programmer p has competences in Java programing is represented as shown in Fig. 4.

Fig. 4. Competence representation example

However, the simple CG model does not allow to adequately represent entities and their competences. Indeed, in a simple CG model, the semantic of a concept type or a relation type is only given by its position in the type hierarchies;

the only mechanism that enables defining a type is the specialization/generalization relation. This representation of types is poor and misses a lot of expressivity to represent generic information about types and also some relation properties such as transitivity, symmetry and reflexivity. To deal with these problems, we propose to use CG rules as described hereafter.

1. *Concept type definition:* To represent generic information about concept types, these types must be defined. "Concept type definition" is defined here as "an either necessary or necessary and sufficient conditions that entities must verify in order to belong to a concept type". These conditions are formalized using conceptual graph rules. For example, the concept type *Mother* defined as a "woman that is *mother_of* a person" is defined as follows:

$$[Mother : *x] \Rightarrow [Woman :?x] \to (mother\_of) \to$$
$$[Person : *]$$
$$Woman : *x] \to (mother\_of) \to [Person : *] \Rightarrow$$
$$[Mother :?x].$$

2. *Relation type definition:* In the same way, a "Relation type definition" is "an either necessary or necessary and sufficient conditions which must be verified in order to belong to a relation type". For example, the relation type *grandmother_of* can be defined as follows:

$$[Woman : *x] \to (grandmother\_of]) \to [Person : \; *y]$$
$$\Rightarrow [Woman : *x] \to (mother\_of) \to [Person : *]$$
$$\to (parent\_of) \to [Person :?y].$$
$$[Woman : *x] \to (mother\_of]) \to [Person : *y] \to$$
$$(parent\_of) \to [Person : *y] \Rightarrow [Woman :?x]$$
$$\to (grandmother\_of) \to [Person :?y].$$

3. *Meta-knowledge on relations:* Relation properties are also formalized using CG rules. For example, the following rules enables expressing the fact that the relations *parent_of* and *child_of* are symmetric ones:

(1) $[Person : *x] \to (child\_of) \to [Person : *y] \Rightarrow$
   $[Person : *y] \to (child\_of) \to [Person :?x].$
(2) $[Person : *x] \to (parent\_of) \to [Person : *y] \Rightarrow$
   $[Person :?y] \to (parent\_of) \to [Person :?x].$

As a result of the rule-based representation we propose, the domain representation is composed of (1) *Ontological knowledge*, represented by the support, to which we add a component named "Rule base" (RB) containing the set of rules used to define the types and the relation properties and, (2) *Factual knowledge*, represented by CGs labeled from the support. In this work, CGs serve for representing entities together with their acquired competences. Each graph is then published in one of the mediators of the federation. The set of the competences that are published in a given mediator are collected into a single CG named "Competence Base" and denoted as CB.

A CB is built and updated every time where a new competence (represented by a CG noted P) is published. For each published graph P, we follow the three following steps:

(1) Disjoint sum of the graphs P and CB in order to add published competences to the CB.
(2) Normalize the graph CB: this normalization avoids graph redundancy and then minimizes the search space.
(3) Apply the rules that are present in RB on the graph CB. This is a very important step: it allows reasoning over the CB in order to add all implicit knowledge that is not directly published into the CB.

3.3 Competence Discovery

This section presents the way a request is represented as a CG, together with a request satisfaction process.

Request Representation. The request is represented as a headed CG form noted RG in which:

(i) The searched entities are represented by the head t of RG. We introduce a special marker ? logically equivalent to the * marker in order to indicate such a node.
(ii) The requested competences are represented by relations which are directly attached to the node t.
(iii) The rest of RG represents conditions on the requested competences.

For example, seeking for men having some competences in UML (Unified Modeling Language) and some competences in programing using languages that support classes is represented as shown in Fig. 5.

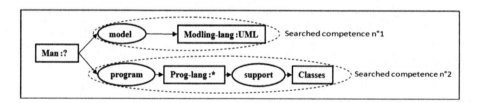

Fig. 5. A request example

Local Request Satisfaction. The local satisfaction of a request R runs as follows: *(i)* Normalize R in order to minimize its size and as a consequence to minimize the search and to avoid logical and semantic ambiguities, *(ii)* delete from R all the connected components that do not contain the head t, because these components are independent from the searched entities, *(iii)* Project R on BC and *(iv)* if at least one projection is found, then there is at least one single answer to the request. Answers are then all the projections (images) of the head node t. Otherwise, search for possible composite answers to the request.

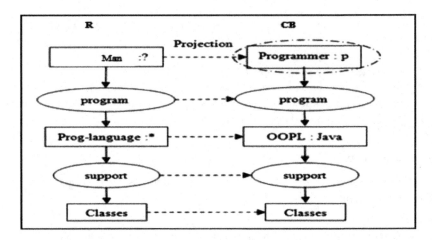

Fig. 6. A local request satisfaction example

As an example, the satisfaction of the request R in the left part in Fig. 6 is the circled concept, the right part of the figure being the concept base.

In order to find possible composite answers to a request R, we decompose R into sub-requests where every sub-request consists in searching entities having one of the required competences and we proceed as follows:

(i) Decompose R into n sub-requests $R_i(i \in [1, n])$, each R_i containing the head of R connected to one of the sub-graphs representing a discovery request for one competence, together with conditions on it (see Sect. 3.3).
(ii) Satisfy all the sub-requests, one independently from the others.
(iii) If all the sub-requests are satisfied then composite answers are the compositions of the answers of the sub-requests.

As an example, to find composite answers to the request in Fig. 5, R is decomposed into two sub-requests (Fig. 7).

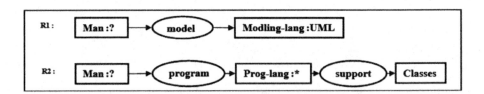

Fig. 7. A request decomposition example

In addition, the satisfaction of a sub-request R_i proceeds as follows:

(i) Project R_i on BC.
(ii) If at least one projection is found then R_i is locally satisfied and the replies to R_i are images of the head node.

(iii) Otherwise, try the distributed satisfaction of R_i thanks to the cooperation with other mediators and this is explained in the coming section.

Cooperative Request Satisfaction. In a federation of mediators, part of a sub-request R_i may be satisfied in one of the mediators of the federation whereas another part may be satisfied in another one. In term of conceptual graphs, this means that a part of the graph that represents R_i may be projected on the CB in one mediator whereas another part may be projected on the CB in another one, as illustrated in the following example, considering the sub-request R2 in Fig. 7.

Assume that two mediators M1 and M2 are available (Fig. 8 shows parts of their competence bases denoted CB1 and CB2 respectively).

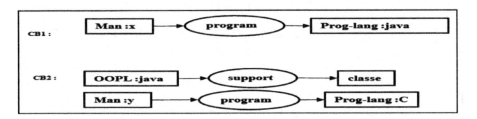

Fig. 8. Competence base examples

In both the mediators, only a part of R_2 is satisfied: in CB1, there is a person having java-programming competences and in CB2, we know that java supports classes. So, in order to satisfy a sub-request in a federation, it is sufficient to find which parts of R_i can be projected on the CB of a mediator and which parts cannot. However, the projection operation such as defined in the CG formalism does not allow to find this type of information. For that reason we propose to proceed according to the following steps:

Step1: Decompose R_i into elementary parts containing only one relation. For example, the sub-request R_2 in Fig. 7 is decomposed into two parts (Fig. 9).

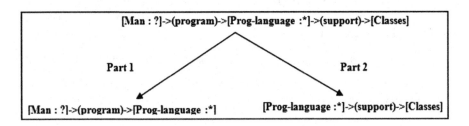

Fig. 9. Sub-request decomposition example.

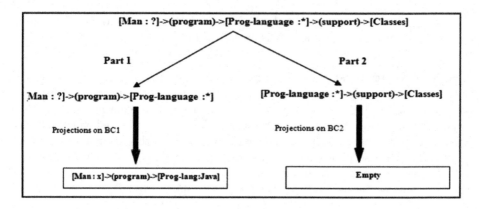

Fig. 10. Sub-request parts projection on CB1

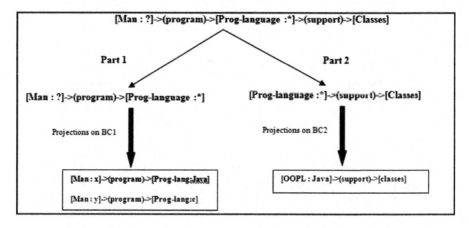

Fig. 11. Sub-request parts projection on CB2

Step2: Project these parts on each CB in the federated mediators:

(1) The projection of the two parts on CB1 is shown in Fig. 10. (2) Add the projection of the two parts on CB2 (Fig. 11).

Step3: Check whether the projections can be joined and if they do, then the sub-request is satisfied and the satisfactions are the projections of the sub-requests' heads (see the dotted parts in Fig. 12).

Let us now describe the prototype we developed as a support of our proposals.

4   Implementation

For an experimental validation of the proposed approach, we implemented a prototype using many software components. There exist several tools which

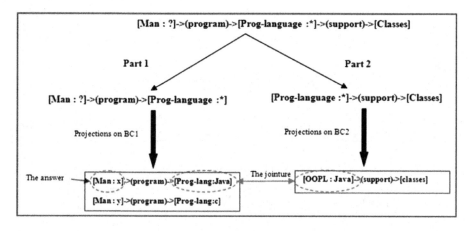

Fig. 12. Sub-request satisfaction verification

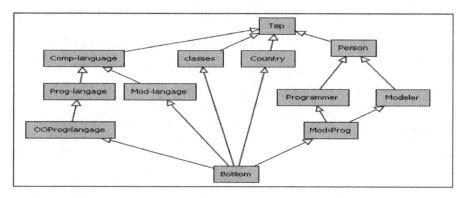

Fig. 13. The concept hierarchy

implement CGs in particular for research purposes and for information extraction [7,12,13,26]. However, few of these tools offer a complete software environment for the widest possible use of the model: the storage and the manipulation of a large number of graphs. For that reason, we choose to use the CoGITaNT library (Conceptual Graphs Integrated Tools allowing Nested Typed graphs), a library of C++ classes (open source, developed at LIRM Montpellier, CNRS, France) which allows developing applications based on the CG knowledge representation scheme.

We illustrate hereafter the prototype functioning thanks to examples. We present first, the domain population and then examples of competence discovery.

(1) **Domain population:** as an example, we consider the computer science competence management domain represented in terms of a concept type hierarchy (Fig. 13), a relation type hierarchy (Fig. 14), rules used to define concept types, relation types and relations properties.

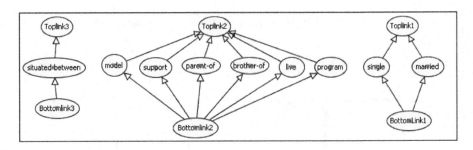

Fig. 14. The relation hierarchy

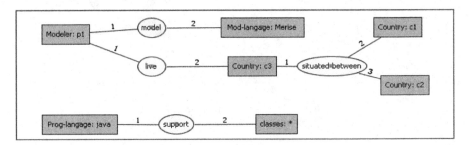

Fig. 15. Competence base of the mediator M1.

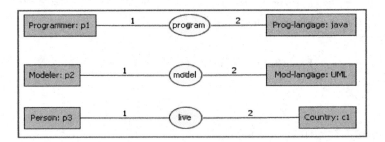

Fig. 16. Competence base of the mediator M2.

(2) **Competence bases** are in the Figs. 15 and 16.
(3) **A local query-satisfaction example** is shown in the Fig. 17, in which *graph2* denotes the query.
(4) **A distributed query-satisfaction example:** the Figs. 18 and 19 illustrate the result of the query denoted as *graph1* in the Fig. 18.

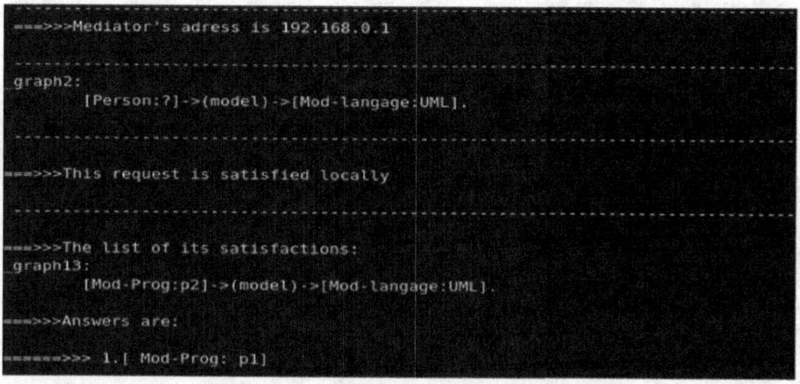

Fig. 17. Local satisfaction of a request

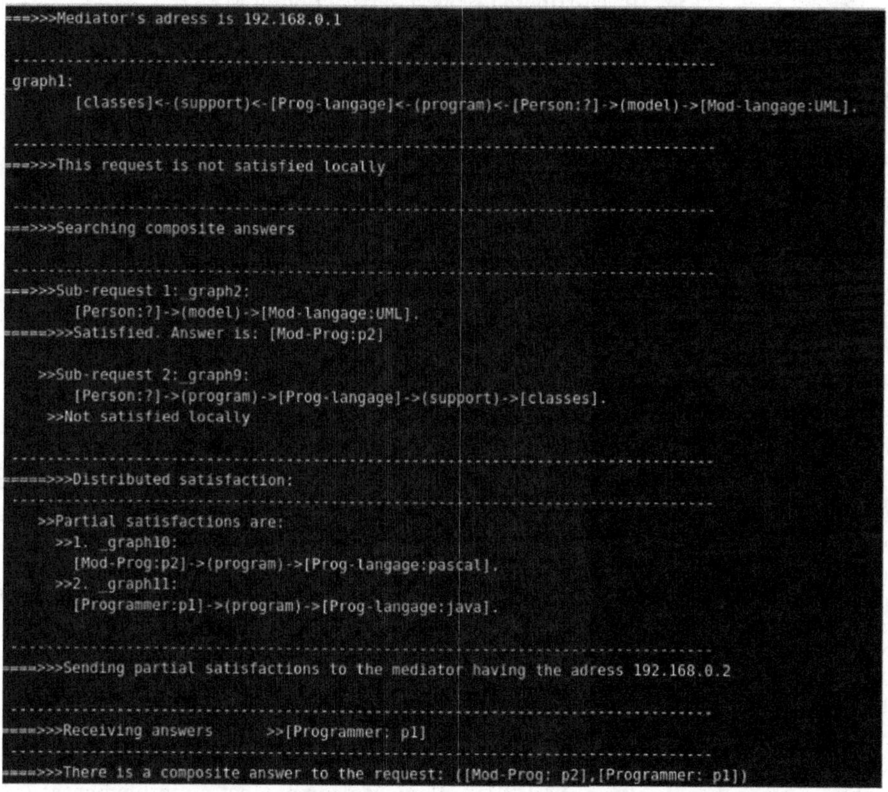

Fig. 18. Distributed request satisfaction in M1.

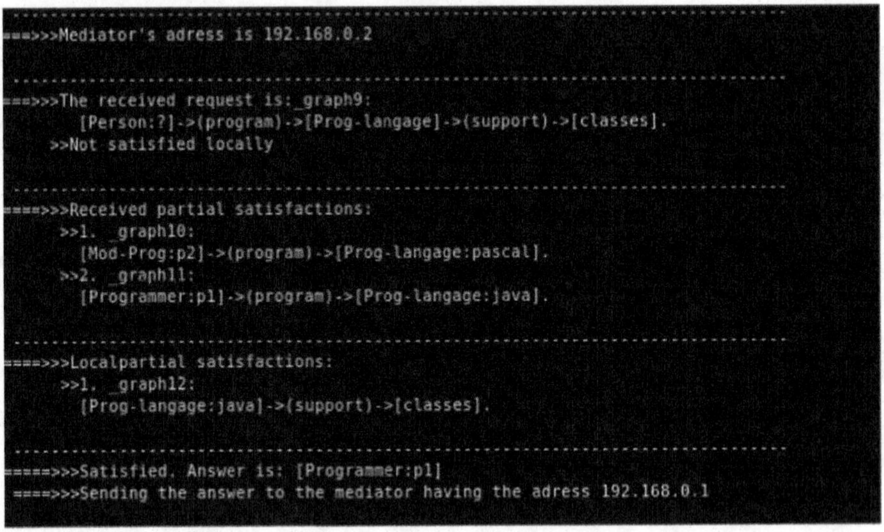

```
===>>>Mediator's adress is 192.168.0.2

===>>>The received request is:_graph9:
       [Person:?]->(program)->[Prog-langage]->(support)->[classes].
    >>Not satisfied locally

===>>>Received partial satisfactions:
    >>1. _graph10:
       [Mod-Prog:p2]->(program)->[Prog-langage:pascal].
    >>2. _graph11:
       [Programmer:p1]->(program)->[Prog-langage:java].

===>>>Localpartial satisfactions:
    >>1. _graph12:
       [Prog-langage:java]->(support)->[classes].

=====>>>Satisfied. Answer is: [Programmer:p1]
=====>>>Sending the answer to the mediator having the adress 192.168.0.1
```

Fig. 19. Distributed request satisfaction in M2.

5   Discussion and Concluding Remarks

In this paper, we presented an approach for competence management and discovery using conceptual graphs (CG) to provide a formal semantic description of an application domain. Acquired competences are organized under a CG form that is built and updated every time a new competence is published. The advantage of this organization form is that the application of graph rules at publication time facilitates the search and may reduce the response time, since all implicit information are available thanks to the application of these rules at publication time. For competence discovery, we use operations on graphs and the projection is used as a basic operation in the discovery process. For distributed satisfaction of a request, we use another form of graph decomposition where a sub-request is decomposed into elementary parts containing only one relation. In addition, for experimentation purposes, we implemented a federated mediation prototype based on the client/server architecture of COGITANT [12]. The prototype is fully written in C++ programming language and it has been successfully verified under Linux and MICROSOFT Windows XP operating systems.

Further work is to consider the complexity of the search algorithm and to cope with heterogeneous mediators cooperation, i.e. mediators where knowledge bases are described in different languages. An additional on-going research topic concerns the dynamic and semantic-based identification of possible cooperating mediators for unsatisfied parts of a competence request together with a performance comparative analysis of a P2P implementation against an implementation using cloud computing technology.

References

1. Berio, G., Harzallah, M.: Knowledge management for competence management. J. Univ. Knowl. Manag., 21–28 (2005)
2. Bernstein, P.A., Giunchiglia, F., Kementsietsidis, A., et al.: Data management for peer-to-peer computing: A vision. In: Proceedings of the 5th International Workshop on the Web and Databases (WebDB), pp. 89–94, June 2002
3. Borgida, A., Devanbu, P.: Adding more dl to idl: towards more knowledgeable component interoperability. In: Proceedings of the 21st International Conference on Software Engineering (ICSE), CA, USA, pp. 378–387, May 1999
4. Bouchikhi, M., Boudjlida, N.: Using larch to specify the behavior of objects in open distributed environments. In: Proceedings of the 5th Maghrebian Conference on Software Engineering and Artificial Intelligence (MCSEAI), Tunis, TUNISIA, pp. 275–287, December 1998
5. Boudjlida, N.: A Mediator-based architecture for capability management. In: Proceedings of the 6th IASTED International Conference on Software Engineering and Applications (SEA), Cambridge, USA, pp. 45–50, November 2002
6. Boudjlida, N., Cheng, D.: Complement concept and capability discovery. In: Proceedings of the 1rst International Workshop on Enterprise Modelling and Ontologies for Interoperability (EMOI) in Connection with the 16th Conference on Advanced Information Systems Engineering (CAiSE), Riga, Latvia, pp. 337–342, June 2004
7. CharGer: Editeur de graphes conceptuels. http://sourceforge.net/projects/charger/
8. Chein, M., Mugnier, M.L.: Graph-Based Knowledge Representation: Computational Foundations of Conceptual Graphs. Springer, London (2008)
9. Cheng, D.: Competence Management and Discovery in Heterogeneous Environments (Gestion et découverte de compétences dans des environnements hétérogènes). Ph.D. thesis, Henri Poincaré-Nancy1 University, France (2008)
10. Cheng, D., Boudjlida, N.: An architecture for heterogenous federated mediators. In: Proceedings of the 17th Conference on Advanced Information Systems Engineering (CAISE), June 2005
11. Cheng, D., Boudjlida, N.: Capability management and discovery in description logic. In: Proceedings of the 3rd International Workshop on Enterprise Modelling and Ontologies for Interoperability (EMOI), in Connection with the 18th Conference on Advanced Information Systems Engineering (CAiSE), pp. 678–681, June 2006
12. CoGITaNT: Conceptual graphs integrated tools allowing nested typed graphs. http://cogitant.sourceforge.net
13. Cogui: Outil de création graphique de graphes conceptuels. http://www.lirmm.fr/cogui
14. Sowa, J.F.: Conceptual Structures: Information Processing in Mind and Machine. Addison-Wesley, USA (1984)
15. Moore, M., Suda, T.: Adaptable peer-to-peer discovery of objects that match multiple keywords. In: International Symposium on Applications and the Internet : Workshops (SAINT Workshops 2004), Tokyo, Japan, pp. 402–407 (2004)
16. Nicolov, N., Mellish, C., Ritchie, G.: Sentence generation from conceptual graphs. In: Ellis, G., Rich, W., Levinson, R., Sowa, J.F. (eds.) ICCS 1995. LNCS, vol. 954, pp. 74–88. Springer, Heidelberg (1995)

17. Nonaka, I., Takeuchi, H.: The Knowledge-Creating Company: How Japanese Companies Create the Dynamics of Innovation. Oxford University Press, New York (1995)
18. Papakonstantinou, Y., Garcia-molina, H., Ullman, J.: Medmaker: A mediation system based on declarative specifications. In: Proceedings of the 12th International Conference on Data Engineering (ICDE), pp. 132–141, February 1996
19. Preece, A.: A mediator-based infrastructure for virtual organisations. In: Agents-2001 Workshop on Intelligent Agents in B2B E-Commerce, Montreal, Canada (2001)
20. Quillian, M.R.: Semantic memory. In: Minsky, M. (ed.) Semantic Information Processing, pp. 227–270. MIT Press, Cambridge (1968)
21. Ram, S., Park, J.S., Hwang, Y.: CREAM: a mediator based environment for modeling and accessing distributed information on the web. In: Eaglestone, B., North, S.C., Poulovassilis, A. (eds.) BNCOD 2002. LNCS, vol. 2405, pp. 58–61. Springer, Heidelberg (2002)
22. Salvat, E.: Raisonner avec des opérations de graphes: graphes conceptuels et règles d'inférence. Ph.D. thesis, Montpellier II University (1997)
23. Schmidt-Schauss, M., Smolka, G.: Attributive concept descriptions with complements. Artif. Intell. J. **48**, 1–26 (1991)
24. Site, C.: Conceptual Graphs Interchange Format. http://www.webkb.org/doc/CGs.html
25. Site, D.: Description logics. http://www.dl.kr.org/
26. Southey, F., Linders, J.G.: Notio - a java API for developing CG tools. In: Tepfenhart, W.M. (ed.) ICCS 1999. LNCS, vol. 1640, pp. 262–271. Springer, Heidelberg (1999)
27. Talantikite, H., Aissani, D., Boudjlida, N.: Semantic annotations for web services discovery and composition. Comput. Stand. Interfaces **31**, 1108–1117 (2009)
28. Tönshoff, H.K., Seilonen, I., Teunis, G., Leito, P. : A mediator-based approach for decentralised production planning, scheduling and monitoring. In: Proceedings of the CIRP International Seminar on Intelligent Computation in Manufacturing Engineering (ICME), June 2000
29. Wiederhold, G.: Mediators in the architecture of future information systems. IEEE Comput. **25**, 38–49 (1992)

Subjective Networks: Perspectives and Challenges

Magdalena Ivanovska[1]([✉]), Audun Jøsang[1], Lance Kaplan[2], and Francesco Sambo[3]

[1] University of Oslo, Oslo, Norway
magdalei@ifi.uio.no, audun.josang@mn.uio.no
[2] US Army Research Lab, Adelphi, MD, USA
lance.m.kaplan.civ@mail.mil
[3] University of Padova, Padova, Italy
francesco.sambo@dei.unipd.it

Abstract. Subjective logic is a formalism for reasoning under uncertain probabilistic information, with an explicit treatment of the uncertainty about the probability distributions. We introduce *subjective networks* as graph-based structures that generalize Bayesian networks to the theory of subjective logic. We discuss the perspectives of the subjective networks representation and the challenges of reasoning with them.

Keywords: Subjective logic · Bayesian networks · Subjective networks

1 Introduction

Subjective logic [5] is a formalism for reasoning under uncertain probabilistic information. The basic entities in subjective logic are *subjective opinions* on random variables. A subjective opinion includes a *belief mass distribution* over the states of the variable, complemented with an *uncertainty mass*, which together reflect a current analysis of the probability distribution of the variable by an expert, based on a test, etc.; and a *base rate* probability distribution of the variable, reflecting a domain knowledge that is relevant to the current analysis. A subjective opinion can always be *projected* onto a single probability distribution, but this necessarily removes information about the uncertainty mass.

While the probability distribution of a random variable represents uncertainty about the value of the variable, a subjective opinion on the variable represents a second-order uncertainty, i.e. uncertainty about the probability distribution itself. The latter is further formalized by establishing a correspondence between subjective opinions and Dirichlet probability density functions [3]. Through this correspondence, the projected probability distribution of the subjective opinion corresponds to expected probabilities or the mean probability values.

Conditional reasoning with subjective opinions has been explored for the case of two variables, resulting in the definition of *deduction* and *abduction* operations

M. Croitoru et al. (Eds.): GKR 2015, LNAI 9501, pp. 107–124, 2015.
DOI: 10.1007/978-3-319-28702-7_7

for multinomial variables [6]. An alternative approach to deduction based on the correspondence with the multinomial Dirichlet model is explored in [8].

This paper attempts to address the conditional reasoning with subjective opinions in general, introducing *subjective networks* as graph-based structures that generalize Bayesian networks to the theory of subjective logic.

A Bayesian network [10] is a compact representation of a joint probability distribution of a set of random variables in the form of directed acyclic graph and a set of conditional probability distributions associated with each node of the graph. The goal of inference in Bayesian networks is to derive the conditional probability distribution of any set of (target) variables in the network, given that the values of any other set of (evidence) variables have been observed. Bayesian networks reasoning algorithms provide a way to propagate the probabilistic information through the graph, from the evidence to the target. Bayesian networks are a powerful tool for modelling and inference of various situations involving probabilistic information about a set of variables, and thus form a base for developing tools with applications in many areas like medical diagnostics, risk management, etc.

One serious limitation of the Bayesian networks reasoning algorithms is that all the input conditional probabilities must be assigned precise values in order for the inference algorithms to work and the model to be analysed. This is problematic in situations where probabilities can not be reliably elicited and one needs to do inference with uncertain or incomplete probabilistic information, inferring the most accurate conclusions possible. Subjective opinions can represent uncertain probabilistic information of any kind (minor or major imprecision, and even total ignorance), by varying the uncertainty mass between 0 and 1.

A straightforward generalization of Bayesian networks in subjective logic retains the network structure and replaces conditional probability distributions with conditional subjective opinions at every node of the network. We call this a *Bayesian subjective network* and consider the reasoning in it a generalization of the classical Bayesian reasoning, where the goal is to obtain a subjective opinion on the target given the evidence. The evidence in this case can be an instantiation of values, but also a subjective opinion itself. In most of the cases, the inference in Bayesian subjective networks remains a challenge, since subjective opinions do not enjoy all the nice properties of the probability distributions; in particular, the notions of conditional, marginal and joint subjective opinion do not have the same interrelations as the corresponding notions in probability theory.

In this paper, we also discuss representation and inference with another type of subjective networks that we call *fused subjective networks*, where the graph follows the available input information as associated with the arrows rather than the nodes, and where information coming from multiple paths to the same node is combined by a *fusion* operation. We give an example of modelling with subjective networks thorough the special case of the *naïve Bayes subjective network*, which can be considered to belong to both the Bayesian and the fused subjective networks type.

The paper is structured as follows: In Sect. 2 we first review the necessary preliminaries from probability theory and Bayesian networks. Then we introduce subjective opinions on random variables and their correspondence with the multinomial Dirichlet model. Section 3 introduces subjective networks representation. In Sect. 4 we introduce the types of inference problems that can be distinguished in subjective networks and discuss potential solutions. Section 5 presents an alternative approach to inference in subjective networks that builds upon the Dirichlet representation of subjective opinions. In Sect. 6 we conclude the paper.

2 Preliminaries

2.1 Bayesian Networks

We assume a simplified definition of *random variable* as a variable that takes its values with certain probabilities. More formally, let X be a variable with a *domain* (set of values, states of the variable) $\mathbb{X}$. A *probability distribution* p of X is a function $p : \mathbb{X} \to [0,1]$, such that:

$$\sum_{x \in \mathbb{X}} p(x) = 1. \tag{1}$$

$p(x)$ is the probability that the variable X takes the value x.

Let $V = \{X_1, \ldots, X_n\}$ be the set of all random variables that are of interest in a given context. A *joint probability distribution* of the variables in V is a probability distribution defined on the Cartesian product of $\mathbb{X}_1, \ldots, \mathbb{X}_n$:

$$\sum_{x_1 \in \mathbb{X}_1} \cdots \sum_{x_n \in \mathbb{X}_n} p(x_1, \ldots, x_n) = 1. \tag{2}$$

In general we talk about sets of variables, subsets of V. A set of variables $Y = \{Y_1, \ldots, Y_k\} \subseteq V$ can also be considered a variable with a domain $\mathbb{Y} = \mathbb{Y}_1 \times \cdots \times \mathbb{Y}_k$. As standard in Bayesian networks literature, we use the notation of a variable also for a set of variables, making the obvious identifications (see [10]).

Given a joint probability distribution p of the variables in V, and a set of variables $Y \subset V$, the *marginal probability distribution* of Y is a function $p : \mathbb{Y} \to [0,1]$ defined by:

$$p(y) = \sum_{x \in \mathbb{X}, \ X = V \setminus Y} p(y, x). \tag{3}$$

Given two sets of variables X and Y, a *conditional probability distribution* of Y given that X takes the value x, $p(Y|x)$, is a function from $\mathbb{Y}$ to $[0,1]$ defined by the following equation:

$$p(y|x) = \frac{p(y, x)}{p(x)}. \tag{4}$$

$p(y|x)$ is the conditional probability that Y takes the value y, given that the value of X is x.

A set of variables X is conditionally independent of a set of variables Y given the set of variables Z, denoted by $I(X, Y|Z)$, if the following holds:

$$p(x|y, z) = p(x|z) \text{ whenever } p(y, z) > 0, \tag{5}$$

for every choice of assignments x, y, and z.

A Bayesian network [10] with n variables is a directed acyclic graph (DAG) with random variables $X_1, \ldots, X_n$ as nodes, and a set of conditional probability distributions $p(X_i|Pa(X_i))$ associated with each node X_i containing one conditional probability distribution $p(X_i|pa(X_i))$ of X_i for every assignment of values $pa(X_i)$ to its parent nodes $Pa(X_i)$. If we assume that the Markov property holds: Every node is conditionally independent on its non-descendants given its parents,

$$I(X_i, ND(X_i)|Pa(X_i)), \tag{6}$$

for the given DAG and the joint distribution p, then p is determined by:

$$p(x_1, \ldots, x_n) = \prod_{i=1}^{n} p(x_i|pa(X_i)), \tag{7}$$

where $pa(X_i)$ is the instantiation of the parents of X_i that corresponds to the tuple $(x_1, \ldots, x_n)$.

The general inference goal in Bayesian networks is to derive the probability $p(y|x)$, for every instantiations x and y of subsets X and Y of V, in an efficient way compatible with the network's topology.

2.2   Subjective Opinions

In this section we review the basic notions related to multinomial and hyper subjective opinions on random variables.

Let X be a random variable. A *multinomial subjective opinion* on X [6] is a tuple:

$$\omega_X = (b_X, u_X, a_X), \tag{8}$$

where $b_X : \mathbb{X} \rightarrow [0, 1]$ is a *belief mass distribution*, $u_X \in [0, 1]$ is an *uncertainty mass*, and $a_X : \mathbb{X} \rightarrow [0, 1]$ is a *base rate distribution*, satisfying the following additivity constraints:

$$u_X + \sum_{x \in \mathbb{X}} b_X(x) = 1, \tag{9}$$

$$\sum_{x \in \mathbb{X}} a_X(x) = 1. \tag{10}$$

The beliefs and the uncertainty mass reflect the results of a current analysis of the random variable applying experts' knowledge, experiments, or a combination of the two. $b_X(x)$ is the belief that X takes the value x expressed as a degree in $[0,1]$. It represents the amount of experimental or analytical evidence in favour of x. u_X is a single value, representing the degree of uncertainty about the distribution of X. It represents lack of evidence due to lack of knowledge or expertise, or insufficient experimental analysis. The base rate a_X is simply a probability distribution of X that represents domain knowledge relevant to the current analysis.

For example, a GP wants to determine whether a patient suffers from depression through a series of different tests. Based on the test results, the GP concludes that the collected evidence is 10 % inconclusive, but is still two times more in support of the diagnosis that the patient suffers from depression than of the opposite one. As a result, the GP assigns 0.6 belief mass to the diagnosis that the patient suffers from depression and 0.3 belief mass to the opposite diagnosis, complemented by 0.1 uncertainty mass. The probability that a random person in the population suffers from depression is 5 % and this fact determines the base rate distribution in the GPs subjective opinion on the condition of the patient.

In some cases of modelling it is useful to be able to distribute belief mass to subsets of $\mathbb{X}$ as well. This leads to generalization of multinomial subjective opinions to *hyper opinions*, which distribute the belief mass over the reduced power set of $\mathbb{X}$ (*hyperdomain of X*), $\mathcal{R}(\mathbb{X}) = \mathcal{P}(\mathbb{X}) \backslash \{\mathbb{X}, \emptyset\}$:

$$b_X : \mathcal{R}(\mathbb{X}) \to [0,1], \tag{11}$$

and u_X is a value from $[0,1]$, such that the following holds:

$$u_X + \sum_{x \in \mathcal{R}(\mathbb{X})} b_X(x) = 1. \tag{12}$$

a_X is again a probability distribution of X, defined on $\mathbb{X}$.[1]

$b_X(x)$ represents the belief that the value of X is (in the set) $x \in \mathcal{R}(\mathbb{X})$, and represents the amount of evidence that supports exactly x.[2]

A subjective opinion in which $u_X = 0$, i.e. an opinion without uncertainty mass, is called a *dogmatic opinion*. Dogmatic multinomial opinions correspond to probability distributions. A dogmatic opinion for which $b_X(x) = 1$, for some $x \in \mathbb{X}$, is called an *absolute opinion*. Absolute multinomial opinions correspond to instantiating values of variables. In contrast, an opinion for which $u_X = 1$, and consequently $b_X(x) = 0$, for every $x \in \mathcal{R}(\mathbb{X})$, i.e. an opinion with complete uncertainty, is called a *vacuous opinion*. Vacuous opinions correspond to complete ignorance about the probability distribution of the variable.

A multinomial opinion ω_X is "projected" to a probability distribution $\mathrm{P}_{\omega_X} : \mathbb{X} \to [0,1]$ defined in the following way:

$$\mathrm{P}_{\omega_X}(x) = b_X(x) + a_X(x)\, u_X. \tag{13}$$

---

[1] For simplicity, we make an abuse of the notation using the same type of letters for both elements of $\mathbb{X}$ and elements of $\mathcal{R}(\mathbb{X})$.

[2] If we think of u_X as of an amount of evidence assigned to the whole domain $\mathbb{X}$, then b_X and u_X correspond to a *basic belief assignment* [12]. However, u_X is a measure for lack of evidence, not a belief, as will be further clarified in the next section.

We call the function P_{ω_X} a *projected probability distribution* or a *projection* of the subjective opinion ω_X. According to Eq. (13), the projected probability $P_{\omega_X}(x)$ is equal to the belief mass in support of x increased by the portion of the base rate of x determined by the uncertainty mass u_X. In this way, the projected probability of x is an estimate for the actual probability of x that varies from the base rate value, in the case of complete ignorance in the current analysis, to the actual probability in the case of zero uncertainty mass.

We call *focal elements* the elements of $\mathcal{R}(\mathbb{X})$ that are assigned a non-zero belief mass. In the case of a hyper opinion there can be focal elements that have a non-empty intersection. For hyper opinions the definition of projected probability distribution is generalized as follows:

$$P_{\omega_X}(x) = \sum_{x' \in \mathcal{R}(\mathbb{X})} a_X(x|x') \, b_X(x') + a_X(x) \, u_X, \tag{14}$$

for $x \in \mathbb{X}$, where $a_X(x|x')$ is a conditional probability of x given x', if a_X is extended to $\mathcal{P}(X)$ additively.[3] If we denote the sum in Eq. (14) by b'_X:

$$b'_X(x) = \sum_{x' \in \mathcal{R}(\mathbb{X})} a_X(x|x') \, b_X(x'), \tag{15}$$

it is easy to check that $b'_X : \mathbb{X} \to [0,1]$, together with u_X, satisfies the additivity property in Eq. (12), i.e. $\omega'_X = (b'_X, u_X, a_X)$ is a multinomial opinion. From Eqs. (14) and (15) we obtain $P_{\omega_X} = P_{\omega'_X}$. This means that every hyper opinion can be approximated with a multinomial opinion which has the same projected probability distribution as the initial hyper one.

2.3   Subjective Opinions as Probability Density Functions

In this section we describe the correspondence between multinomial opinions and multinomial Dirichlet models given in [3].

Let $\boldsymbol{p} = (p_1, \ldots, p_k)$ be the probability distribution of the variable X, where $p_i = p(x_i)$, $i = 1, \ldots, k$. $\boldsymbol{p}$ is Dirichlet distributed random variable if its probability density function (pdf) has the following form:

$$f_{\boldsymbol{\alpha}}(\boldsymbol{p}) = \frac{\Gamma\left(\sum_{i=1}^{k} \alpha_i\right)}{\prod_{i=1}^{k} \Gamma(\alpha_i)} \prod_{i=1}^{k} p_i^{\alpha_i - 1}, \tag{16}$$

where Γ is the k-dimensional ($k = |\mathbb{X}|$) gamma function and $\boldsymbol{\alpha} = (\alpha_1, \ldots, \alpha_k)$ are the parameters of the distribution. The *mean* distribution determined by Eq. (16) is given as $m(x_i) = \alpha_i / \sum_{i=1}^{k} \alpha_i$.

The *multinomial Dirichlet model* [1] for the probability distribution $\boldsymbol{p}$ assumes the following:

---

[3] For this conditional probability to be always defined, it is enough to assume $a_X(x_i) > 0$, for every $x_i \in \mathbb{X}$. This amounts to assuming that everything we include in the domain has a non-zero probability of occurrence.

- a *Dirichlet prior* pdf for $\boldsymbol{p}$ with parameters $\alpha_i^0 = Ca_X(x_i)$, where a_X is the mean distribution and C is a constant called *prior strength* that determines the amount of evidence needed to overcome the prior, and
- a *multinomial sampling* $(r(x_i) \mid i = 1, \ldots, k)$, $N = \sum_{i=1}^{k} r(x_k)$, i.e. N observations where x_i is observed $r(x_i)$ times.

Then the posterior pdf for $\boldsymbol{p}$ is also a Dirchlet pdf with the following parameters:

$$\alpha_i = r(x_i) + C \, a_X(x_i), \tag{17}$$

and a corresponding mean distribution:

$$m(x_i) = \frac{r(x_i) + Ca_X(x_i)}{N + C}. \tag{18}$$

The posterior Dirichlet pdf for $\boldsymbol{p}$ uniquely determines a multinomial opinion $\omega_X = (b_X, u_X, a_X)$, where a_X is the prior mean distribution and the belief and uncertainty masses are determined in the following way:

$$\begin{cases} b_X(x_i) = \frac{r(x_i)}{N+C} \\ u_X = \frac{C}{N+C} \end{cases} \tag{19}$$

According to the above transformation, the beliefs are proportional to the corresponding observations while the uncertainty is independent of the particular observations and decreases with their total number. When the total number of observations converges to infinity, the uncertainty converges to zero and the beliefs converge to the actual $\boldsymbol{p}$, i.e. the transformation determines a dogmatic opinion. The projected probability of the opinion ω_X obtained by the transformation in Eq. (19) is equal to the mean value of the posterior Dirichlet pdf as given in Eq. (18), which corresponds to the fact that the projected probability is an estimate for the actual distribution $\boldsymbol{p}$ of X.

Conversely, given subjective opinion $\omega_X = (b_X, u_X, a_X)$ and Dirichlet strength C determine Dirichlet parameters $\boldsymbol{\alpha}$ as given in Eq. (17), where:

$$r(x_i) = \frac{Cb_X(x_i)}{u_X}. \tag{20}$$

The correspondence described in this section implies that expressing the knowledge about X in the form of subjective opinion $\omega_X = (b_X, u_X, a_X)$ is equivalent to expressing the knowledge about the probability distribution of X by a multinomial Dirichlet pdf. This correspondence gives a way of eliciting the beliefs and uncertainty in subjective opinions from experimental analysis. However, the base rate as well as its strength C have to be chosen in advance.

3 Subjective Networks Representation

A *subjective network* S_n of n random variables is a directed acyclic graph and sets of subjective opinions associated with it. First we introduce the concepts of joint and conditional subjective opinion, then we introduce two different types of subjective networks, Bayesian and fused subjective networks.

3.1   Conditional and Joint Subjective Opinions

A *joint subjective opinion* on variables $X_1, \ldots, X_n, n \geq 2$ is a tuple:

$$\omega_{X_1 \ldots X_n} = (b_{X_1 \ldots X_n}, u_{X_1 \ldots X_n}, a_{X_1 \ldots X_n}), \tag{21}$$

where $b_{X_1 \ldots X_n} : \mathcal{R}(\mathbb{X}_1 \times \ldots \times \mathbb{X}_n) \rightarrow [0, 1]$ and $u_{X_1 \ldots X_n} \in [0, 1]$ satisfy the additivity condition in Eq. (12) and $a_{X_1 \ldots X_n}$ is a joint probability distribution of the variables $X_1, \ldots, X_n$.

A *marginal opinion* on a set of variables Y, subset of $V = \{X_1, \ldots, X_n\}$, is a joint opinion on the variables in Y. The relation between a marginal opinion on the variables Y and a joint opinion on the full set of variables V can not be modelled with an analogue of Eq. (3), but rather with what is known as *product operation* [3] in subjective logic, where a product of two multinomial opinions on independent random variables is defined as a joint hyper opinion on the Cartesian product of their domains. The definition is generalizable to an arbitrary number of variables and to opinions on sets of variables, under the assumption that the input opinions are subjective opinions on probabilistically independent (sets of) variables.

Given two sets of random variables X and Y, a *conditional opinion* on Y given that X takes the value x is a subjective opinion on Y defined as a tuple:

$$\omega_{Y|x} = (b_{Y|x}, u_{Y|x}, a_{Y|x}), \tag{22}$$

where $b_{Y|x} : \mathcal{R}(\mathbb{Y}) \rightarrow [0, 1]$ and $u_{Y|x} \in [0, 1]$ satisfy the condition in Eq. (12) and $a_{Y|x} : \mathbb{Y} \rightarrow [0, 1]$ is a probability distribution of Y. We use the notation $\omega_{Y|X}$ for a set of conditional opinions on Y, one for each value of X, i.e.:

$$\omega_{Y|X} = \{\omega_{Y|x} \mid x \in X\}. \tag{23}$$

There is no relation in subjective logic analogous to Eq. (4) that relates conditional opinions with the marginal ones.

3.2   Bayesian Subjective Networks

A *Bayesian subjective network* of n random variables $X_1, \ldots, X_n$ is a directed acyclic graph with one node for each variable and a set of conditional subjective opinions $\omega_{X_i|Pa(X_i)}$, associated with each node X_i, consisting of one conditional opinion $\omega_{X_i|pa(X_i)}$ on X_i, for each instantiation $pa(X_i)$ of its parent nodes $Pa(X_i)$.

A Bayesian subjective network is basically a generalization of a classical Bayesian network, where instead of probability distributions associated with the nodes, we have subjective opinions on them. Conversely, every Bayesian subjective network projects to a classical one. Namely, every opinion $\omega_{X_i|pa(X_i)} \in \omega_{X_i|Pa(X_i)}$ projects to a probability distribution $P(X_i|pa(X_i))$. The graph of the given subjective network S_n together with the sets of projected distributions $P(X_i|Pa(X_i))$, $i = 1, \ldots, n$, forms a classical Bayesian network, which we denote by $P(S_n)$ and call a *Bayesian network projection* of the subjective network S_n.

The concepts of joint, marginal, and conditional opinions do not enjoy the same relations as their probabilistic counterparts. Consequently, the joint opinion on $X_1, \ldots, X_n$ can not be obtained from the input opinions in the network using the Markov condition by an analogue of Eq. (7). Nevertheless, the corresponding projected probabilities are related by the equations in Sect. 2.1 and can be reasoned about within the Bayesian network $P(S_n)$.

The Bayesian subjective networks representation also presumes that Markov independences, hence all the conditional independences embedded in the graph structure of the given DAG (d-separations), hold for the uncertainties of the corresponding opinions: If a set of variables X is conditionally independent of a set of variables Y given the set of variables Z, $I(X, Y|Z)$ then:

$$u(X|yz) = u(X|z), \tag{24}$$

for every choice of values y and z. This assumption can be justified by the fact that the uncertainty mass of a subjective opinion is a parameter that refers to the whole probability distribution. In light of the Dirichlet pdf representation of subjective opinions, a subjective network represents in some sense an ensemble of possible Bayesian networks, where the spread of the distributions is related to the uncertainties. For each ensemble, $I(X, Y|Z)$ implies $p(X|yz) = p(X|z)$. Therefore, the spread of $p(X|yz)$ is the same as that of $p(X|z)$.

A subjective network S_n is a graphical representation of uncertain information about probability distributions that combines beliefs and uncertainty, as well as probabilistic information about the knowledge domain in the form of base rate distributions. The base rate distributions in the conditional opinions of a subjective network S_n can be set without constraints and are not necessarily connected by the equations in Sect. 2.1, i.e. the subjective network may or may not represent a joint probability distribution on the knowledge domain. In the two-node case considered in [6], it is assumed $a_{Y|x_i} = a_Y$, for every $x_i \in X$, i.e. that only the unconditional base rate distributions are available.

Example: Naïve Bayes Subjective Networks. A specific case which is often modelled in data mining and machine learning is that of a set of variables $X_1, \ldots, X_n$, all conditionally independent given another variable Y, so that the joint distribution of the $n + 1$ variables can be decomposed as follows:

$$p(x_1, \ldots, x_n, y) = p(y) \prod_i p(x_i|y). \tag{25}$$

The relations between the variables can be represented with what is known as a *naïve Bayes* network, where Y is the common root node with $X_1, \ldots, X_n$ as children. Such a model is amenable for its scarcity of parameters, compared to the full joint distribution, and for the possibility of assessing each $p(X_i|y)$ independently of the others, possibly from different sources of information or at different times.

Having uncertain information about the probability distributions $p(X_i|Y)$ and $p(Y)$ in the form of subjective opinions, we obtain a *naïve Bayes subjective network* (Fig. 1).

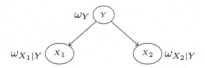

Fig. 1. A three-node Naïve Bayes subjective network

As an example, let us suppose we want to construct a subjective network for detecting type 2 diabetes (T2D), which will be the common root node, from four of its major risk factors, which will be the children: obesity, old age, family history of T2D, and past episodes of high blood glucose. In this case, the choice of the naïve Bayes network structure is practical: information on T2D prevalence and on the probability distribution of the children nodes with and without diabetes is easy to gather from the appropriate medical sources, whereas information on the joint distribution of the four variables would be much harder to get. Then, in constructing the input opinions $\omega_{X_i|y} = (b_{X_i|y}, u_{X_i|y}, a_{X_i|y})$, the uncertainty mass could be set higher on the conditional opinions when y is *true* and lower when y is *false*, on the account of the much larger amount of samples from which the latter probabilities are probably estimated. Furthermore, uncertainty mass could be set higher on the opinions on family history and past episodes of high blood glucose than on the ones on age and obesity, on the account of the latter two being more reliable to assess precisely and not being based on the memory of past events or on historical clinical records. In the lack of clear evidence for T2D in a particular case, the opinion ω_Y could be set to vacuous, where the only relevant information we use is the domain knowledge (statistics on T2D in the population, for example) reflected in the base rate distribution a_Y.

3.3   Fused Subjective Networks

A *fused subjective network* is a DAG and sets of conditional subjective opinions $\omega_{Y|X}$ as defined in Eq. (23), for every pair of nodes X and Y, such that X is a parent of Y, i.e. a DAG and a set of conditional subjective opinions at each of its arrows. In addition, the fused subjective networks representation assumes that base rates for the root nodes of the DAG are also available.

This second type of subjective networks resembles more closely what has originally been introduced in [6]. The purpose of such a representation is to facilitate the elicitation of experts' opinions on variables that form a graph containing V-structures, the simplest one given in Fig. 2. In such a graph, it can be easier for the analyst to provide the opinions $\omega_{Y|X_1}$ and $\omega_{Y|X_2}$ separately, rather than the opinions $\omega_{Y|X_1X_2}$, which is otherwise necessary if we want to construct a Bayesian subjective network with the same DAG (Fig. 3).

For example, an expert might have an opinion on the probability of military aggression over a country A from a country B conditional on the troop movements in B, and opinion on the probability of military aggression conditional on the political relations between the countries, but is not able to merge these

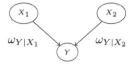

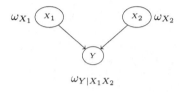

Fig. 2. A V-structure fused subjective network

Fig. 3. A V-structure Bayesian subjective network

opinions in a single opinion about military aggression conditional on both the factors considered together.

Unlike in Bayesian subjective networks, the projected probability distributions of the subjective opinions in a fused subjective network do not necessarily form a Bayesian network with the given DAG, which is a substantial difference between the two representations that strongly influences the inference approach. A fused subjective network is an approximation of the Bayesian subjective network with the same DAG, in terms of both representation and inference.

Note that in naïve Bayes networks and, in general, subjective networks with a DAG that is a single-rooted tree, every node has at most one parent, hence this type of subjective networks is in the intersection of fused and Bayesian ones.

4   Inference in Subjective Networks

The inference in classical Bayesian networks reduces to the following: Given that the values of some of the variables (evidence variables E) are observed, to find the probability of any other variable or set of variables in the network (target variables T), which is to find the conditional probability of the target given an instantiation of the evidence, $p(T|e)$. In subjective networks, the evidence does not necessarily mean an observation. Namely, an analogue to observing the value of a variable in the case of subjective opinions is assigning a belief mass 1 to that particular value of the variable (based on direct observation, or just a strong opinion). This gives an evidence in the form of an absolute opinion. In general, we could have evidence in the form of a general type of subjective opinion on E, ω_E, and would like to be able to account for it, i.e. to be able to update the opinions on the target variables upon the evidence ω_E.

In subjective networks we can distinguish among three different types of subjective evidence:

- *absolute evidence*, evidence in the form of an absolute opinion, i.e. instantiation of the evidence variables,
- *dogmatic evidence*, evidence in the form of a dogmatic opinion on the evidence variables, and
- *uncertain evidence*, evidence in the form of a subjective opinion with uncertainty greater than zero.

For the derived opinion on the target T we use the notation $\omega_{T\|e}$ in the case of absolute evidence, and $\omega_{T\|E}$ in the case of dogmatic or uncertain evidence. Depending on whether E is a set of one or more variables, we distinguish between:

- *single evidence*, evidence on one variable only, and
- *multiple evidence*, evidence on more than one variable.

4.1   Inference in a Two-Node Network

In this section we briefly summarize the operations of deduction and abduction for conditional reasoning with two variables defined in [4,6]. We assume we have a two-node subjective network, where X is the parent and Y is the child node, and the set of conditional subjective opinions $\omega_{Y|X} = \{\omega_{Y|x} \mid x \in X\}$ is available, along with the base rate distributions a_X and a_Y.

Subjective Logic Deduction. Given the set of opinions $\omega_{Y|X}$ and a subjective evidence ω_X, the goal is to *deduce* a subjective opinion on Y, $\omega_{Y\|X}$[4].
First the projected probability distribution of $\omega_{Y\|X}$ is determined by:[5]

$$P(y\|X) = \sum_{x \in X} P(x)P(y|x). \qquad (26)$$

For the belief masses of the deduced opinion $\omega_{Y\|X}$, we assume the following: The unconditional beliefs of the deduced opinion are at least as large as the minimum of the conditional beliefs:

$$b_{y\|X} \geq \min_{x \in X}\{b_{y|x}\}, \qquad (27)$$

for every $y \in Y$. This is a natural assumption, which can also be found as a *principle of plausible reasoning* for example in [11]. Then we first determine the uncertainty mass $u_{Y\|\hat{X}}$ corresponding to a vacuous evidence opinion on X as the maximum possible uncertainty mass under the conditions in Eqs. (26) and (27). The uncertainty of the deduced opinion from ω_X is then determined as a

---

[4] This input information can be considered both a Bayesian subjective network with no evidence, or a fused subjective network with a subjective evidence on X.

[5] Note that in this section we use simplified notation for the projected probabilities, beliefs and base rates, for example $P(x_i)$ is an abbreviation of $P_{\omega_X}(x_i)$, $b_{y\|X}$ is an abbreviation of $b_{Y\|X}(y)$, etc.

weighted average of the uncertainty based on vacuous input opinion for X and the input conditional uncertainties:

$$u_{Y\|X} = u_X u_{Y\|\hat{X}} + \sum_{x \in X} b_x u_{Y|x}. \tag{28}$$

Once we have the uncertainty mass of the deduced opinion, the beliefs are easily derived using the projected probabilities and Eq. (13).

Subjective Logic Abduction. Given the set of opinions $\omega_{Y|X}$ and an opinion ω_Y, the goal is to *abduce* an opinion on X, denoted by $\omega_{X\|Y}$[6].

The operation of abduction first "inverts" the given set of conditional opinions $\omega_{Y|X}$ into a set of conditional opinions $\omega_{X|Y}$ and then applies deduction on $\omega_{X|Y}$ and ω_Y to derive the opinion $\omega_{X\|Y}$.

For determining the inverted opinion $\omega_{X|y}$, we first obtain its projected probability distribution as follows:

$$P(x|y) = \frac{a_x P(y|x)}{\sum_{x' \in \mathbb{X}} a_{x'} P(y|x')}. \tag{29}$$

Then its uncertainty mass $u_{X|y}$ is obtained by a heuristic procedure which takes the maximum possible uncertainty value compatible with Eq. (29) and adjusts it using the average uncertainty of the input conditionals $\omega_{Y|X}$ and the irrelevance of X to the value y, for details see [4]. The beliefs $b_{x|y}$ are again a consequence of the projected probabilities and the uncertainty.

4.2   Inference in Bayesian Subjective Networks

Let us assume that we are given a Bayesian subjective network S_n and absolute evidence on the set of variables E. Given the instantiation e of E, we want to find the conditional subjective opinion on a target set of variables T, $\omega_{T\|e}$. We make the following general assumptions for the inference procedure for deriving $\omega_{T\|e}$:

1. The projected probability of the derived opinion is determined from the projected Bayesian network in a classical way, i.e. $P(T|e)$ is determined in $P(S_n)$ using the standard Bayesian networks reasoning methods.
2. All the conditional and marginal base rate distributions in the subjective network are either given *a priori*, or determined from the ones provided in the initial opinions in the network by Bayesian reasoning.
3. The uncertainty and the beliefs of the derived opinion satisfy certain constraints, like, for example, some of the conditions given in Eqs. (24) and (27).

The first assumption provides a way of determining the projected probability distribution of the target opinion $\omega_{T\|e}$ and is a starting point in deriving the

---

[6] This case can only be classified as fused subjective network with a subjective evidence on Y.

opinion. Having determined the projected probability and considering the second assumption above, Eq. (13) gives us a system of m linear equations with the beliefs and the uncertainty mass of the target opinion as variables, where m is the cardinality of the domain of the target T. We obtain one more equation over the same variables from the additivity property for the beliefs and uncertainty of subjective opinions, given in Eq. (12). This means that we have a system of $m+1$ equation with $m+1$ variables, which might seem to fully determine the required opinion $\omega_{T\|e}$. However, Eq. (12) is a sum of the equations in Eq. (13), which means that the system is dependent. Hence, the system has infinite number of solutions, i.e. there are infinitely many subjective opinions with the same projected probabilities and base rates, and additional constraints on beliefs and the uncertainty mass are required (assumption 3.) in order to choose a single opinion on the target as a solution.

The above discussion implies the following: If we find a suitable way of determining the uncertainty mass of the derived opinion, the beliefs follow from Eq. (13) (the base rate is either *a priori* given or determined from the given base rates), and the opinion is fully derived. While this is successfully applied in the deduction for two variables described in the previous section, in general, it remains a challenge to provide a meaningful way of propagating the given uncertainty masses throughout the network in a way that would give reasonable belief mass values (that satisfy the initially set constrains) as a consequence. Also, there might not exist a unique way of propagating the uncertainty, and how we decide to do it can be context-dependent.

The above described inference procedure would operate over multinomial opinions. It is possible though to provide the input information in the form of hyper opinions, in which case their multinomial approximations (described at the end of Sect. 2.2) can be used in the inference procedure to derive a multinomial opinion on the target. This is an advantage in a certain sense for one usually has the input information in the more vague, hyper opinion form, and wants to have the output as a multinomial opinion, i.e. to have the beliefs distributed over the values rather than sets of values.

The inference from dogmatic or uncertain evidence remains a challenge, for in that case we can not have the assumption 1, namely: Instantiating the evidence variables E in a given subjective network S_n with a subjective opinion ω_E that is not absolute, we simultaneously provide a new projected probability distribution of E, which, in general, differs from the one that would be derived by Bayesian reasoning in $P(S_n)$.

4.3   Inference in Fused Subjective Networks

We limit the inference in fused subjective networks to the case of a single target variable, i.e. we define the inference problem as follows: Given subjective opinions on the evidence variables $X_1, \ldots, X_k$, derive a subjective opinion on the target variable $Y \notin \{X_1, \ldots, X_k\}$.

Consider first the simple case of a fused subjective network with a singly-connected DAG (only one path between any two nodes) and a simple inference problem where we have a single evidence: Given a subjective opinion ω_X on a node X, derive a subjective opinion $\omega_{Y\|X}$ on another node Y. This problem can be solved by propagating the evidence through the path from X to Y by applying deduction or abduction operation at each step (depending on the direction of the arrows on the path). If the graph is not singly-connected and there are multiple paths between X and Y, then this procedure will enable us to derive multiple different opinions on the target given the same evidence. An operation of *fusion* in subjective logic can then be applied to fuse the derived opinions on the target into a single one.

An operation of fusion can also be applied in the case of multiple evidence: We derive a subjective opinion on the target variable for each of the evidence opinions separately, and then fuse the resulting opinions to obtain a single one. However, in a general graph structure, paths between evidence and target variables can intersect and partially overlap, and to avoid repetitions in the inference procedure, we should consider fusing the opinions on the same variable coming from different paths before propagating them further.

There is a variety of fusion operators [2] that can be used for fusion in subjective networks, hence choosing an appropriate one is one of the challenges in the inference in fused subjective networks.

In some cases, inference in a fused subjective network can be done by first transforming it into a Bayesian one in the following way: for every V-structure with parents $X_1, \ldots, X_n$ and child Y, we invert the given set of conditionals opinions $\omega_{Y|X_i}, i = 1, \ldots, n$ into $\omega_{X_i|Y}$ as described in the abduction operation in Sect. 4.1. This means that we invert the V-structure into a naïve Bayes network where Y is a parent of $X_1, \ldots, X_n$. Because of the conditional independences in the naïve Bayes, we can apply the product operation from [3] on the opinions $\omega_{X_i|y}, i = 1, \ldots, n$ to obtain the opinion $\omega_{X_1 \ldots X_n|y}$, for every $y \in \mathbb{Y}$, i.e. the set of opinions $\omega_{X_1 \ldots X_n|Y}$. At the end, we invert again to obtain the set $\omega_{Y|X_1 \ldots X_n}$.

5 Inference Through the Dirichlet Representation of Subjective Opinions

This section provides an alternative approach towards inference in subjective networks, which is based on the Dirichlet pdf representation of subjective opinions introduced in Sect. 2.3.

In a subjective network, evidence has been collected to form subjective opinions about the conditional probabilities. In other words, each conditional probability distribution $p(X_i|pa(X_i))$ is represented as a k_i-dimensional Dirichlet distributed random variable, where $k_i = |\mathbb{X}_i|$. Because of the Markov property, these Dirichlet distributed random variables are also statistically independent.

One important goal of inference from absolute evidence in subjective networks is to derive an opinion $\omega_{X_i\|e}$ for a given instantiation e of evidence variables E, and a target node X_i not in E. In terms of the Dirichlet representation, determining $\omega_{X_i\|e}$ is equivalent to determining the appropriate Dirichlet pdf to represent the uncertainty about the probability distribution $p(X_i|e)$. According to Sect. 2.1, this probability distribution is expressed through the input probability distributions in the graph in the following way:

$$p(x_i|e) = \frac{\sum_{X_j \in V \setminus E \cup \{X_i\}} \prod_{k=1}^{n} p(x_k|pa(X_k))}{\sum_{X_j \in V \setminus E} \prod_{k=1}^{n} p(x_k|pa(X_k))}, \tag{30}$$

where $pa(X_k)$ is the instantiation of the parents of X_k that corresponds to x_i and e.

For a standard Bayesian network, the execution of Eq. (30) can be accomplished as a series of variable elimination steps [14]. For subjective networks, the probability distributions involved in the right-hand side of Eq. (30) are Dirichlet distributed random variables and exact inference becomes more challenging. The target probability distribution $p(X_i|e)$ is a k_i-dimensional random variable characterized through the independent Dirichlet distributed random variables $p(X_k|pa(X_k))$. Through a change of variables process, it is possible to determine the actual pdf for $p(X_i|e)$, which in general is not a Dirichlet pdf.

In order to obtain a subjective opinion on X_i given e by means of the transformation in Eq. (19), we need to approximate this pdf by a Dirichlet pdf. We choose to use a moment matching approach to determine the best Dirichlet pdf to approximate the pdf of $p(X_i|e)$. First, the mean value of this Dirichlet pdf, $m(X_i|e)$, must equal the expectation of the actual pdf for $p(X_i|e)$. Then, the Dirichlet strength is selected so that the second order moments of the actual target distribution matches that of the Dirichlet distribution as much as possible in the mean squared sense. The matching of the second order moments is perfect only for binary variables. The moment matching method to determine the Dirichlet strength has been implemented for partial observation updates and deduction in [7,8], respectively. In the general case where the evidence can come from the descendants of X_i, a closed form solution for the first and second order expectation of Eq. (30) does not exist because of its fractional form, and one must resort to numerical integration over N Dirichlet distributed random variables, where N is the number of input probability distributions in the network. Such a moment matching method is only computationally feasible for the smallest of networks.

Current research is looking at extending the sum-product algorithm [13]. Such an approach develops a divide and conquer strategy that will provide means to propagate one piece of evidence at a time. Then the effects of an observation coming from the antecedents is propagated forward via subjective logic deduction (as in [8]), and a backwards process will enable the computation of the effect of an observation coming from a descendant node. At each stage in the process, the stored conditionals are approximated by Dirichlet distributions using the moment matching method. Finally, the inference of the target opinion from combined evidence is accomplished by normalizing the opinions conditional on evidence coming from different directions. The first steps of this normalization process has been studied in [9] for the case of a three-node chain of binary variables.

The evaluation of forward/backward propagation along with normalization over chains is the next step. The intermediate results will be stored as subjective opinions, which means that the inference via normalization will only be an approximation of moment matching of Eq. (30), which is not making any Dirichlet approximation about the marginal distribution for the intermediate nodes between the evidence E and X_j. This is in contrast to the sum-product algorithm over Bayesian networks, which provides exact inference. The plan is to study computational efficiency and accuracy of imposing the Dirichlet approximation as the effects of the observations propagate over the "uncertain" probabilistic edges.

The development and evaluation of inference techniques over subjective networks will consider increasing complexity in various dimensions. One dimension is the topology of the network, where we will first study chains and then expand to trees and eventually arbitrary DAGs where we will need to modify the sum-product framework. Another dimension is the complexity of the subjective opinions: We start with binary $(k_i = 2)$ and multinomial opinions $(k_i > 2)$, to finally consider hyper opinions $(2^{k_i} - 2)$. The quality of the observations over E provides another complexity dimension to explore. Initially, we will only consider inference from absolute opinions, which are equivalent to instantiation of variables, but in future work we plan to consider inference from general type of opinions.

6   Conclusions and Future Work

We introduced subjective networks as structures for conditional reasoning with uncertain probabilistic information represented in the form of subjective opinions on random variables. In this way both the input information and the inferred conclusions in the modelled scenario incorporate a current analysis of beliefs and a domain knowledge, at the same time taking the uncertainty about the probabilities explicitly into account.

The discussed inference problems in subjective networks lead to several inference approaches to be studied in future work: global uncertainty propagation in Bayesian subjective networks, piece-wise inference in fused subjective networks, and a statistical moment matching approach towards inference in subjective networks.

Acknowledgments. The research presented in this paper is conducted within the project ABRI (Advanced Belief Reasoning in Intelligence), partially funded by the US Army Research Laboratory.

References

1. Gelman, A., Carlin, J., Stern, H., Dunson, D., Vehtari, A., Rubin, D.: Bayesian Data Analysis, 2nd edn. Chapman and Hall/CRC, Florida (2004)
2. Jøsang, A., Hankin, R.: Interpretation and fusion of hyper opinions in subjective logic. In: Proceedings of the 15th International Conference on Information Fusion (FUSION 2012) (2012)

3. Jøsang, A., McAnally, D.: Multiplication and comultiplication of beliefs. Int. J. Approx. Reason. **38**(1), 19–51 (2004)
4. Jøsang, A., Sambo, F.: Inverting conditional opinions in subjective logic. In: Proceedings of the 20th International Conference on Soft Computing (MENDEL 2014), Brno (2014)
5. Jøsang, A.: A logic for uncertain probabilities. Int. J. Uncertainty, Fuzziness Knowl.-Based Syst. **9**(3), 279–311 (2001)
6. Jøsang, A.: Conditional reasoning with subjective logic. J. Multiple-Valued Logic Soft Comput. **15**(1), 5–38 (2008)
7. Kaplan, L., Sensoy, M., Chakraborty, S., de Mel, G.: Prtial observable update for subjective logic and its application for trust estimation. Inf. Fusion **26**, 66–83 (2015)
8. Kaplan, L., Sensoy, M., Tang, Y., Chakraborty, S., Bisdikian, C., de Mel, G.: Reasoning under uncertainty: variations of subjective logic deduction. In: Proceedings of the 16th International Conference on Information Fusion, FUSION 2013, Istanbul (2013)
9. Kaplan, L., Ivanovska, M., Jøsang, A., Sambo, F.: Towards subjective networks: extending conditional reasoning in subjective logic. In: Proceedings of the 18th International Conference on Information Fusion, FUSION 2015, Washington, DC (2015)
10. Pearl, J.: Probabilistic Reasoning in Intelligent Systems. Morgan Kaufman Publishers, San Francisco (1988)
11. Pearl, J.: Reasoning with belief functions: an analysis of compatibility. Int. J. Approx. Reason. **4**(5–6), 363–389 (1990)
12. Shafer, G.: A Mathematical Theory of Evidence. Princeton University Press, Princeton (1976)
13. Wainwright, M.J., Jordan, M.I.: Graphical models, exponential families, and variational inference. Found. Trends Mach. Learn. **1**(1), 1–305 (2008)
14. Zhang, N.L., Poole, D.: A simple approach to bayesian network computations. In: Proceedings of the Tenth Canadian Conference on Artificial Intelligence (1994)

RDF-SQ: Mixing Parallel and Sequential Computation for Top-Down OWL RL Inference

Jacopo Urbani[1,2($\boxtimes$)] and Ceriel Jacobs[1]

[1] Department of Computer Science, VU University Amsterdam,
Amsterdam, The Netherlands
{jacopo,ceriel}@cs.vu.nl
[2] Max Planck Institute for Informatics, Saarbruecken, Germany

Abstract. The size and growth rate of the Semantic Web call for querying and reasoning methods that can be applied over very large amounts of data. In this paper, we discuss how we can enrich the results of queries by performing rule-based reasoning in a top-down fashion over large RDF knowledge bases.

This paper focuses on the technical challenges involved in the top-down evaluation of the reasoning rules. First, we discuss the application of well-known algorithms in the QSQ family, and analyze their advantages and drawbacks. Then, we present a new algorithm, called RDF-SQ, which re-uses different features of the QSQ algorithms and introduces some novelties that target the execution of the OWL-RL rules.

We implemented our algorithm inside the QueryPIE prototype and tested its performance against QSQ-R, which is the most popular QSQ algorithm, and a parallel variant of it, which is the current state-of-the-art in terms of scalability. We used a large LUBM dataset with ten billion triples, and our tests show that RDF-SQ is significantly faster and more efficient than the competitors in almost all cases.

1 Introduction

The ability to derive implicit and potentially unknown information from graph-like RDF datasets [8] is a key feature of the Semantic Web. This process, informally referred to as reasoning, can be performed in several ways and for different purposes. In this paper, we focus on the application of reasoning to enrich the results of SPARQL queries [14] by deriving implicit triples that are relevant for the query, and restrict our focus to rule-based reasoning in the OWL 2 RL fragment.

In this context, one method to perform reasoning is traditionally referred to as *backward-chaining*. The main idea behind backward-chaining is to rewrite the input query in a number of subqueries whose results can be used by the rules to calculate implicit answers. Backward-chaining is often implemented with a *top-down* evaluation of the rules, where the "top" is the input query and the "bottom" consists of the queries that cannot be rewritten.

We consider the top-down algorithms designed for the Datalog language [2] because almost all rules in OWL 2 RL can be represented with this language. In

© Springer International Publishing Switzerland 2015
M. Croitoru et al. (Eds.): GKR 2015, LNAI 9501, pp. 125–138, 2015.
DOI: 10.1007/978-3-319-28702-7_8

Datalog, the most popular algorithms in this category belong to the QSQ family [2,4], which consists of a series of algorithms that implement the well-known SLD-resolution technique [9]. These algorithms differ from each other because their computation can be either recursive or iterative, tuple or set oriented, sequential or parallel, or with or without adornments.

Regardless of the chosen algorithm, a common problem of backward-chaining is that reasoning is performed under tight time constraints (since typically the user waits until the query is computed) and the computation might become too expensive to guarantee an acceptable response time. On the Web, this problem is worsened by the fact that the size of current RDF datasets increases continuously. Therefore, it is paramount to develop scalable inference techniques to provide answers to the user in a timely manner.

To address this problem, we studied the salient characteristics of the existing QSQ algorithms, and designed a new algorithm in this family – which we call *RDF-SQ* – that is tailored to the characteristics of the OWL rules and RDF datasets. This algorithm contains several novelties: First, it exploits a pre-materialization of the terminological knowledge and uses it to divide the OWL rules in four different categories depending on the number and type of literals in their body. Each category is implemented in a different way, exploiting the pre-materialization and some heuristics that hold on current RDF datasets. Second, it introduces a new rules evaluation order that interleaves the execution of rules of the first two categories with rules of the last two. The first two categories of rules are executed in parallel, with the goal of collecting as much inference as possible, while the other two are executed sequentially. In doing so, our algorithm interleaves parallel and sequential computation in order to achieve higher efficiency and better utilization of modern hardware.

We tested the performance of our implementation against QSQ-R, the most well-known implementation, and a parallel variant of it that was recently applied over very large RDF knowledge bases. We used as a test ruleset a large fragment of the OWL-RL rules. Our experiments show that RDF-SQ outperforms both algorithms significantly, allowing in this way the execution of SPARQL queries with complex inference over very large knowledge graphs with more than ten billion triples.

2   Background

We assume a basic familiarity with the RDF data model [8]. Typically, users query RDF graphs using the SPARQL language [14], which can be seen as a SQL-like language to retrieve and process sub-portions of the RDF graphs.

SPARQL is a complex and rich language, but every SPARQL query can be represented at its core as a graph pattern, and its execution can be translated into a graph matching problem [14]. In this paper, we consider the most popular type of SPARQL queries, which are the ones that can be mapped with *basic graph patterns (BGP)*. These graph patterns are simply defined as a finite set of *triple patterns*, which are members of the set $(T \cup V) \times (I \cup V) \times (T \cup V)$, where T is a finite set of RDF terms, V of variables, and $I \subseteq T$ is the set of IRIs.

We use the Datalog language to formalize the process of inferring new information from the input. Due to space constraints we only briefly introduce key concepts in this language and refer the reader to [2] for a complete overview. A generic Datalog rule is of the form $R_1(w_1) \leftarrow R_2(w_2), R_3(w_3), ..., R(w_n)$. The left-hand side of the arrow is called the *head* of the rule, while the right-hand side constitutes the rule's *body*. We call each $R_x(w_x)$ with $x \in \{1..n\}$ a *literal*, which is composed of a *predicate* (R_x) and a tuple of terms $w_x := t_1, ..., t_m$. Predicates can be either *intensional* (*idb*) or *extensional* (*edb*), and only intensional predicates can occur in the head of a rule. Each Datalog *term* can be either a variable or a constant (in this paper, variables are always indicated with capital letters to distinguish them from the constants). We call a literal a *fact* if the tuple contains only constants. We say that a fact f *instantiates* a literal l if every constant in l is equal to the constant at the same position in f, and there is an unique mapping between each variable in l and a corresponding constant at the same position in f. Consequently, the instantiation $f \leftarrow f_1, f_2, \ldots, f_n$ of a rule is a sequence of facts where the mapping from constants to variables is unique across the entire rule.

In Datalog, instantiations of rules are typically calculated through the manipulation of *substitutions*, which map variables to either constants or other variables and are calculated using special functions (called θ in this work). Sets of substitutions can be joined together ($\bowtie$) or retrieved from a database of facts using a generic function called *lookup*. An *unifier* is a special substitution between two literals that is often used to verify whether the head of a rule can produce instantiations for a given literal. A unifier that is no more restrictive than needed is called a most general unifier (*MGU*). In this work, we use the usual definitions of these concepts. Their formal definition, and all other Datalog concepts not explicitly defined in this paper, can be found in [2,4,21].

Given a generic database I which contains a finite set of Datalog facts and a ruleset R, we say that a fact f is an *immediate consequence* of I and R if either $f \in I$ or there exists an instantiation $f \leftarrow f_1, f_2, \ldots, f_n$ of a rule in R where all f_i are in I. Calculating all immediate consequences with a rule r is a process that we refer to as the evaluation of rule r. We define T_R as a generic operator that calculates all immediate consequences so that $T_R(I)$ contains all immediate consequences of I and R. Let $T_R^0(I) := I$, $T_R^1 = T_R(I)$ and for each $i > 0$ let $T_R^i(I) := T_R(T_R^{i-1}(I))$. Since T_R is a monotonic operator, and no new symbol is generated, there will be an i where $T_R^i(I) = T_R^{i-1}(I)$. We call this database the *fixpoint* of I and denote it with $T_R^\omega(I)$.

The goal of our work is to answer SPARQL queries over $T_R^\omega(I)$. To this end, two main techniques are normally adopted: The first technique, called *forward-chaining*, stems from fixpoint semantics and consists of calculating the entire $T_R^\omega(I)$ and then (re)using the extended database to answer the SPARQL query. Forward-chaining is often implemented with a *bottom-up* evaluation of the rules, which consists of a repetitive evaluation of the rules over augmented versions of the database. This technique has been explored extensively in literature and there are several systems that implement this type of reasoning with different degrees of expressivity [7,12,18,19,27–29].

The second technique is called *backward-chaining* (or query rewriting), and is the focus of this work. It adopts a proof-theoretic approach and calculates only the subset of $T_R^\omega(I)$ necessary to answer the query. For the purposes of query answering, backward-chaining is efficient because it does not always calculate the entire derivation like forward-chaining. For this reason, backward-chaining is adopted in large-scale RDF query engines like Virtuoso [6], 4Store [17], QueryPIE [22], or Stardog [15].

Backward-chaining is normally implemented with a *top-down* rules evaluation (a notable exception is represented by the Magic Set technique [3], which can be seen as backward-chaining performed with a bottom-up evaluation). We illustrate the functioning of a typical top-down algorithm with a small example.

Example 1. Suppose that the Datalog query $Q := (A, typ, person)$ must be evaluated using a ruleset R on a database I that contains RDF triples encoded as a ternary relation T. We report the content of I and R below:

<table>
<tr><td colspan="1" align="center">*Database I*</td></tr>
<tr><td>T(a,has_grade,3), T(d,has_grade,null), T(b,has_grade,6),</td></tr>
<tr><td>T(student,sc,scholar) T(c,has_grade,7), T(greater,typ,trans)</td></tr>
<tr><td>T(7,greater,6), T(scholar,sc,person) T(6,greater,3),</td></tr>
<tr><td>T(has_grade,dom,student)</td></tr>
</table>

Ruleset R
$R_1 := T(A, sc, C) \leftarrow T(A, sc, B), T(B, sc, C)$
$R_2 := T(A, typ, C) \leftarrow T(B, sc, C), T(A, typ, B)$
$R_3 := T(A, typ, C) \leftarrow T(P, dom, C), T(A, P, B)$
$R_4 := T(A, P, C) \leftarrow T(P, typ, trans),$
$\qquad\qquad\qquad\quad T(A, P, B), T(B, P, C)$

In general, a top-down algorithm would first identify which rules might produce some answers. In our example, these are R_2, R_3, R_4. Then, it would launch a number of subqueries necessary to execute the rules. In our case, R_2 would require the results of the query $T(B, sc, person)$, R_3 of $T(P, dom, person)$, and R_4 of $T(typ, typ, trans)$. These subqueries might either trigger other rules or return no result. In our example, the first subquery would trigger R_1 which would first read $T(scholar, sc, person)$ and consequently request the subquery $T(B, sc, scholar)$ in order to calculate the triples that instantiate $T(B, sc, person)$. In our case, R_1 would return the fact $T(student, sc, person)$ to R_2, which could use this fact to issue another subquery $T(A, typ, student)$. This last subquery would trigger rule R_3, which would return the facts that a, b, c, d are students back to R_2. At this point, R_2 would use this information to infer that a, b, c, d are of type *person* and return these facts as answers to the original query. $\square$

Unfortunately, a major problem of backward-chaining is that the number of subqueries might become too large to provide all answers in a timely manner. To reduce this issue, backward-chaining algorithms often use *tabling* to reduce the number of evaluations. Tabling is a particular technique of *memoization* [16] and consists of caching the results of previously evaluated queries and reuse them if needed. Tabling can be either *transient* in case it is maintained only during the query execution, or *permanent* if the results are being reused across multiple queries.

3   RDF-SQ

In Datalog, the most popular type of top-down algorithms are the ones in the QSQ family [4]. The most popular QSQ algorithm is called *QSQ-R* and was presented in 1986 [23]. QSQ-R is a recursive sequential algorithm, which exhaustively evaluates each subquery before it returns the results to the rule that generated it. In this way, it can exploit tabling efficiently.

Parallel and distributed versions of QSQ have been proposed in [1,21]. The last contribution is particularly interesting since it weakens the admissibility test and lemma resolution in SLD/AL – that is the theoretical foundation upon which QSQ-R and QoSaQ are based [25,26] – by allowing the rewriting of equivalent queries in case they do not share any parent query except the root. This choice is clearly less efficient than QSQ-R, since the latter does not perform this redundant computation, but it has the advantage that it can run in parallel without expensive synchronization.

In this landscape, our contribution consists of a new algorithm, called RDF-SQ, which is inspired by these methods but is especially designed to execute the OWL RL rules. It introduces novelties that exploit characteristics of current RDF data, and reuses features of existing QSQ algorithms in an intelligent way. We can summarize the following as its salient features:

- Rules are divided into four categories, and the evaluation of rules in each category is implemented with different algorithms;
- Both permanent and transient tabling are used extensively: Terminological knowledge is pre-materialized beforehand as described in [21], and intermediate results are cached in main memory and reused during the process;
- The algorithm interleaves sessions where the rules are evaluated in parallel, and sessions where the rules are evaluated sequentially. This strategy seeks the best compromise between tabling and parallel computation.

In the following we describe RDF-SQ in more detail. First, we describe the categorization of the rules. Then, we give an informal description and report the pseudocode. Finally, we analyze the fundamental properties of termination, soundness, and completeness.

3.1   RDF-SQ: Rule Categories

When the database is loaded, the first operation performed by the system consists of pre-materializing all triples that instantiate a number of predefined *idb* literals using the procedure described in [21]. The goal of this procedure is to avoid a redundant inference upon the same literals during the execution of multiple queries. Therefore, we can see it as a form of permanent tabling: In fact, the results of this pre-materialization are intended to be kept in main memory so that they can be easily retrieved. During the pre-materialization, the original rules are being rewritten by replacing the predicates of the pre-materialized literals with an *edb* predicate that cannot trigger further inference. It has been shown that this rewriting is harmless (after the pre-materialization), and inference using the rewritten rules produces the same derivation(s) as with the original ones [21]. Therefore, our algorithm uses the rewritten rules instead of the original ones.

Before we describe our algorithm, we introduce a categorization of rules into four disjoint categories depending on the number and type of literals they use. We also outline how the categories of rules are implemented in our system, since these two elements are instrumental in understanding the main intuition that motivates our method. In the following, we describe each category in more detail.

Category 1. This category contains all rules that have, as body literals, a fixed number of extensional predicates. These are mostly the pre-materialized literals. An example is R_1 of Example 1. In this case, all triples that instantiate the bodies of these rules are already available in main memory. Therefore, during the rule evaluation we calculate the rule instantiations by performing a number of nested loop joins [5] followed by a final projection to construct the instantiations of the head.

Category 2. This category contains all rules that have as body literals one or more pre-materialized literals and exactly one non-materialized literal. Two examples are rules R_2 and R_3 of Example 1.

These rules are more challenging than the previous ones because their evaluation requires one relational join between a set of tuples that is available in main memory and generic triples that might reside on disk. In our implementation, the pre-materialized triples are indexed in main memory, and a hash-based join is executed as new generic triples are being fetched either from the knowledge base or from other rules.

Categories 3 and 4. The third category contains rules with two or more fixed non-materialized literals (e.g. rule R_4 of Example 1) while the fourth category contains the rules where the number of literals depend on the actual input. These are the ones that use elements of the RDF lists.

These rules are the most challenging to evaluate since they also require joins between two or more generic sets of triples that can reside both on main memory and disk. These joins are performed by first collecting all triples that instantiate the first generic literal in main memory, and then passing all acceptable values to the following generic literal using a sideways-information passing strategy [2]. This process is repeated until all generic patterns are processed. At this point a final projection is used to construct the triples that instantiate the rule's head.

3.2   RDF-SQ: Main Intuition

The system receives in input a collection of triples that constitute the input database, a set of rules, and a SPARQL BGP query. As a first operation, the system performs the pre-materialization and rewrites the initial rules as described in [21]. Then, each triple pattern in the SPARQL BGP query is retrieved in a sequence, and the bindings of each variable are passed to the following pattern using sideways information passing. After the bindings are retrieved, they are joined using an in-memory hash join.

The RDF-SQ algorithm is invoked to retrieve all triples that instantiate each triple pattern. Therefore, we can abstract the inference as a process that takes as input a query that equals to a single literal and rewrites it in multiple subqueries evaluating the rules to produce the derivations.

During this evaluation, a key difference between the categories of rules is that rules in the third and fourth categories need to collect all the results of their subqueries before they can proceed with the rest of the evaluation. If these subqueries trigger further inference, then the rule evaluator must wait until the subqueries are finished. In case the rules are evaluated by different threads, the evaluator must introduce a synchronization barrier to ensure that all subqueries have terminated. In contrast, rules in the first category can be executed independently since their input is already available in main memory, and rules in the second category do not need to wait because they can produce the derivation immediately after they receive one triple. Therefore, the evaluation of the first two categories of rules can be parallelized rather easily, while in the third and fourth categories the synchronization barriers reduce the parallelism.

In RDF-SQ we leverage this distinction and only execute rules of the first two categories in parallel. These rules are executed first, so that we can collect as much derivation as possible before we start to apply the other two rule categories, which require more computation.

3.3   RDF-SQ: Pseudocode

We report the pseudocode of RDF-SQ in Algorithm 1. To perform the parallel evaluation of the rules in the first two categories, we use the parallel version of QSQ presented in [21], which we call *ParQSQ* from now on. In our pseudocode, this algorithm is represented by the function **ParQSQ_infer** and it corresponds to the function "infer" in Algorithm 1 of [21]. For the purpose of this paper, we can see **ParQSQ_infer** as a top-down algorithm that receives in input a query Q and a list of queries already requested (called *SubQueries* in our code) and returns a number of triples that instantiate Q using the ruleset P and $\mathcal{I} \cup Mat$ as input (notice that these variables are marked as global). Internally, this function produces the same computation as in its original version, with the only difference that in the original code *SubQueries* is a local variable, while in our version it is a global synchronized variable, so that every time a new member is added through union (e.g. in line 23), the addition is permanent. This change is necessary to implement our intended behavior of expanding each query only once, like QSQ-R.

Algorithm 1. RDF-SQ Main Algorithm. Q is the input query, $\mathfrak{I}$ is a finite set of facts, and R is the ruleset. The function returns the set of all facts that instantiate Q. $\mathfrak{I}, P, R, Tmp, Mat$ are global variables.

```
 1  function rdf-sq(Q, R, ℑ)
 2    P := R₁₂ := {r ∈ R : r.type = 1 ∨ r.type = 2}
 3    R₃₄ := {r ∈ R : r.type = 3 ∨ r.type = 4}
 4    Mat, Tmp, New := ∅
 5    do
 6      SubQueries := ∅
 7      Mat := Tmp ∪ New ∪ Mat
 8      New := New ∪ ParQSQ_infer(Q, SubQueries)
 9      Mat := Mat ∪ Tmp
10      for(∀SQ ∈ SubQueries ∪ {Q})
11        if SQ was already processed in this loop
12          if all queries in SQ are processed
13            goto line 32
14          else
15            continue
16        else
17          mark SQ as processed
18        all_subst := {θₑ}
19        for ∀r ∈ R₃₄ s.t. SQ is unifiable with r.HEAD
20          θₕ := MGU(SQ, r.HEAD)
21          subst := {θₑ}
22          for ∀p ∈ r.BODY
23            tuples := ParQSQ_infer(θₕ(p), SubQueries ∪ {Q})
24            Tmp := Tmp ∪ tuples
25            subst := subst ⋈ lookup(θₕ(p), tuples)
26          end for
27          all_subst := all_subst ∪ (subst ∘ θₕ)
28        end for
29        if SQ = Q then New := New ∪ ⋃_{θ∈all_subst}{θ(SQ)}
30        else Tmp := Tmp ∪ ⋃_{θ∈all_subst}{θ(SQ)}
31      end for
32    while New ∪ Tmp ⊆ Mat ∪ ℑ
33    return New
34  end function
```

We divide the functioning of RDF-SQ in three steps. First, the rules are divided in two different ruleset categories (lines 2,3). The first ruleset is assigned to P so that it is visible to ParQSQ. Second, the algorithm applies the rules in the first two categories (line 8), and all the derivation produced is collected in Mat. Third, the rules of third and fourth types are applied sequentially on each (sub)subquery produced so far (lines 10–31), and ParQSQ_infer is invoked on each subquery that might be requested by these rules. Notice that the inner invocation of ParQSQ_infer might increase the size of $SubQueries$. Therefore, in order not to enter in a infinite loop we mark each subquery as "processed" and exit after all queries have been processed by all rules. The overall process must be repeated until no rule has derived any new triple (line 32). Finally, the program returns all explicit and inferred triples that instantiate the query Q.

3.4   RDF-SQ: Termination, Soundness, Completeness

In this paper, we limit to discuss these properties only informally since formal proofs are lengthy and can be easily obtained with slight modifications of the proofs presented in [21] for the algorithm ParQSQ.

Termination. In general, every Datalog program is guaranteed to terminate because no new symbol is being introduced [2]. Our algorithm is not an exception: Both the *lookup* and *ParQSQ* functions were proven to terminate [21], and the two loops in the program will eventually terminate because there is always a maximum numbers of facts and queries that we can construct from the domain of a given input database.

Soundness. Soundness is a property that holds if every fact returned by `rdf-sq`(Q, R, I) is contained in $T_R^\omega(I)$ and instantiates Q. In our case, this property can be verified rather easily since the derivations can be generated either by `ParQSQ_infer` or by the retrieval and union of all substitutions in lines 23 and 29. These two operations are equivalent to the operations performed by `ParQSQ_infer` to produce the conclusions. Hence they are sound due to the proof in [21].

Completeness. Completeness requires that every fact that is in $T_R^\omega(I)$ and instantiates Q is returned by `rdf-sq`(Q, R, I). Completeness is a property that has "cursed" QSQ algorithms since their inception. In fact, the original version of QSQ-R presented in [23] was found to be incomplete and was fixed by the same author and others in following publications [13,24]. Despite these fixes, later implementations of QSQ presented in [10,22] and also the widely cited version in [2] are still incomplete. A good explanation for the source of this incompleteness is reported in [11]: Basically, the mistake is in relying on the intuitive assumption that if we re-evaluate a query until fix-point during the recursive process then we eventually retrieve all answers. Unfortunately, there are cases where answers derived in previous steps cannot be exploited by the current subquery because the query is subsumed by a previous subquery, and hence not further expanded.

One solution to fix this problem is to clear the cache of precomputed subqueries either at every recursive call or only on the main loop. In our pseudocode, this operation is performed in line 6 of Algorithm 1. This guarantees that in every iteration all intermediate derivations are used in every rule evaluation, and in every iteration all unique subqueries are fully expanded by every rule at least once. Therefore, our algorithm is complete because the main loop in lines 5–32 will not exit until no more derivation has been produced.

4 Evaluation

To evaluate our contribution, we compared the performance of RDF-SQ against ParQSQ and QSQ-R. We chose the first because it has shown the best scalability [21], and the second because it is the most popular QSQ algorithm. To this end, we implemented both RDF-SQ and QSQ-R algorithms inside the QueryPIE prototype, which contains the original implementation of ParQSQ.

QueryPIE is an on-disk RDF query engine written in Java, which is freely available[1]. It is written on top of Ajira [20] – a general-purpose parallel frame-

---

[1] https://github.com/jrbn/querypie.

Table 1. Response time of the LUBM queries on 10B triples. The numbers in bold represent the best results.

Q	Response time (ms. or seconds if with 's')						Results (#)
	ParQSQ		QSQ-R		RDF-SQ		
	C	W	C	W	C	W	
1	759	**11**	**637**	14	763	**11**	4
3	**1.6 s**	15	1.8 s	41	2.4 s	15	6
4	**4.4 s**	93	5.1s	330	6.6 s	55	34
5	**9.3 s**	251	9.8 s	658	11.0 s	82	719
7	3.5 s	67	4.1 s	236	**2.1 s**	51	4
8	**165.8 s**	3.0 s	176.0	5.3 s	171.8 s	821	7790
10	1.2 s	56	1.2 s	128	**1.1 s**	44	4
11	**9.4 s**	27	9.4 s	35	9.5 s	29	224
12	26.0 s	466	26.7 s	1.2 s	**25.4 s**	238	15
13	-	-	4636.8 s	549.9 s	**4062.3 s**	50.3 s	37118

work for data intensive applications that gives the possibility of splitting computation into concurrent tasks called *chains*. We set up Ajira so that it could launch at most 8 concurrent tasks.

Testbed. We used a machine equipped with a dual quad-core CPU of 2.4 GHz, 24 GB of main memory and an internal storage of two disks of 1 TB in RAID-0. We chose to launch our experiments using the LUBM benchmark for several reasons: (*i*) LUBM is the *de facto* standard for measuring the performance of OWL reasoners over very large RDF datasets; (*ii*) it was recently used to evaluate the performance of state-of-the-art OWL reasoners (e.g. [12,18]); (*iii*) it supports challenging inference that uses rules in all four categories, and contains a representative set of queries that encode different workloads.

To allow a fair comparison between the approaches, we activated the same subset of rules and pre-materialized queries that were used in [21]. The excluded rules are mainly redundant or used to derive a contradiction (these rules cannot be activated during SPARQL answering). The only notable exclusions are the rules that handle the *owl:sameAs* semantics. However, since LUBM does not support this inference, these exclusions does not impact the performance of our implementation.

Query Response Time. We loaded an input dataset that consists of a bit more than 10 billion RDF triples (LUBM(80000)), and launched the LUBM queries with the inference performed by the three top-down algorithms. We report in Table 1 the cold and warm runtimes, and the number of results of each query. Unfortunately not all queries succeeded because QueryPIE requires that all intermediate results must fit in main memory, and this precludes the execution of queries 2, 6, 9, and 14. Also, query 13 failed for ParQSQ: it ran out of memory after about 4 h.

We measured the runtime from the moment that the query is launched to the time where all the data is collected and ready to be returned. The cold runtime reports the runtime after the system is started. The warm runtime consists of the average of the 29 consecutive runs of the same query. These last runs are performed where all the data is in memory and no disk access is performed.

The results presented in the table give some interesting insights. First of all, we notice that the difference between cold and warm reasoning reaches two orders of magnitude. This shows that I/O has a significant impact on the performance.

Second, we notice that RDF-SQ produced the shortest warm runtime in all but one case. To better understand the behaviour, we collected additional statistics during these executions and report them in Table 2. In this table, we report the maximum amount of bytes read from disk, the number of concurrent Ajira chains produced during the execution, and the number of queries requested to the knowledge base.

We chose to record these statistics because the amount of bytes read from disk gives an indication of the I/O cost required for answering the query, while the number of Ajira tasks and queries give a rough indication of the amount of reasoning that was triggered. For example, query 1 is highly selective and triggers no reasoning: In fact, only 21 megabytes are read from disk and the number of both chains and queries is small. The most I/O intensive is query 13, where about 200GB are read from disk.

Looking at the results reported in the two tables, we can draw some further conclusions. First of all, the cold runtime is clearly limited by the I/O speed. All three algorithms read about the same amount of data, except for query 13 where ParQSQ fails. Second, considering the number of chains and queries produced, we notice how ParQSQ is generally inefficient while QSQ-R positions itself as the second most efficient algorithm. However, even though ParQSQ is less efficient than QSQ-R, its runtimes are still competitive since the warm runtime is faster than QSQ-R in almost all cases. This is due to its ability to parallelize the computation.

Performance Breakdown. It is remarkable that RDF-SQ produced the smallest number of subqueries and Ajira tasks in all cases. This highlights the efficiency of RDF-SQ when compared with the other two methods. We further investigated the reasons behind this difference and found that it is due to several factors:

- The impact of parallelism on the warm runtime is limited, since most of the execution time is taken by rules of the third and fourth category. However, parallelism brings a substantial reduction of the cold runtime. For example, the execution of query 13 on a smaller data set (LUBM(8000)) with a single processing thread produces a cold runtime of about 323 s, while if we use two threads the runtime lowers to 124 s. There is no significant difference if we increase the number of threads since two threads are enough to saturate the bandwidth.

Table 2. Statistics for the tests of Table 1. PQ stands for ParQSQ, QR for QSQ-R, while RQ abbreviates RDF-SQ.

Q.	Max MB from disk	# Tasks			# Queries		
		PQ	QR	RQ	PQ	QR	RQ
1	21	18	26	18	6	8	6
3	64	144	266	138	69	98	66
4	216	7086	3275	708	3220	1055	289
5	542	6686	2907	557	2958	990	274
7	473	4527	2635	659	2029	859	282
8	13,846	6467	2427	871	2844	798	370
10	126	4476	1529	610	2006	510	261
11	1,594	57	70	52	20	22	17
12	4,359	4999	3476	775	2257	1132	327
13	198,858	-	3112	555	-	1055	274

– Executing rules of third and fourth category in a sequential manner brings substantial benefits because at every step we can fully exploit tabling and reuse all previous derivations. This is also confirmed by the fact that QSQ-R produces comparable results with ParQSQ despite it being a sequential algorithm while the other is parallel.

5   Conclusions

Overall, our evaluation gives a first empirical evidence of how our strategy of interleaving the execution between two stages, one parallel and one sequential, is beneficial. Because of this, our algorithm produced response times that are significantly lower than other state-of-the-art algorithms using an input with ten billion triples, which can be seen as graphs with more than 10 billion edges.

In the future, we plan to do further experiments to test the performance on larger queries and on datasets with higher expressivity. Furthermore, a promising research direction consists of developing techniques which can dynamically estimate whether parallelism can bring some benefit. Finally, we plan to extend inference to SPARQL queries that encode other types of graph patterns.

To conclude, our contribution shows that complex top-down OWL inference can be applied successfully over very large collections of data. This pushes forward current boundaries, and enables the enrichment of the results of queries over RDF data on an unprecedented scale.

Acknowledgments. This project was partially funded by the COMMIT project, and by the NWO VENI project 639.021.335.

References

1. Abiteboul, S., Abrams, Z., Haar, S., Milo, T.: Diagnosis of asynchronous discrete event systems: datalog to the rescue!. In: Proceedings of the Twenty-Fourth ACM SIGMOD-SIGACT-SIGART Symposium on Principles of Database Systems, pp. 358–367. ACM (2005)
2. Abiteboul, S., Hull, R., Vianu, V.: Foundations of Databases, vol. 8. Addison-Wesley, Reading (1995)
3. Bancilhon, F., Maier, D., Sagiv, Y., Ullman, J.D.: Magic sets and other strange ways to implement logic programs. In: Proceedings of the Fifth ACM SIGACT-SIGMOD Symposium on Principles of Database Systems, pp. 1–15. ACM (1985)
4. Ceri, S., Gottlob, G., Tanca, L.: What you always wanted to know about datalog (and never dared to ask). IEEE Trans. Knowl. Data Eng. **1**(1), 146–166 (1989)
5. DeWitt, D., Naughton, J., Burger, J.: Nested loops revisited. In: Proceedings of the Second International Conference on Parallel and Distributed Information Systems, pp. 230–242. IEEE Comput. Soc. Press (1993)
6. Erling, O., Mikhailov, I.: Virtuoso: RDF support in a native RDBMS. In: de Virgilio, R., Giunchiglia, F., Tanca, L. (eds.) Semantic Web Information Management, pp. 501–519. Springer, Heidelberg (2009)
7. Heino, N., Pan, J.Z.: RDFS reasoning on massively parallel hardware. In: Cudré-Mauroux, P., et al. (eds.) ISWC 2012, Part I. LNCS, vol. 7649, pp. 133–148. Springer, Heidelberg (2012)
8. Klyne, G., Carroll, J.J.: Resource description framework (RDF): Concepts and abstract syntax. W3C recommendation (2006)
9. Lloyd, J.W.: Foundations of Logic Programming. Springer, New York (1984)
10. Madalińska-Bugaj, E., Nguyen, L.A.: Generalizing the QSQR evaluation method for horn knowledge bases. In: Nguyen, N.T., Katarzyniak, R. (eds.) New Challenges in Applied Intelligence Technologies, pp. 145–154. Springer, Heidelberg (2008)
11. Madalińska-Bugaj, E., Nguyen, L.A.: A generalized QSQR evaluation method for horn knowledge bases. ACM Trans. Comput. Logic **13**(4), 32:1–32:28 (2012)
12. Motik, B., Nenov, Y., Piro, R., Horrocks, I., Olteanu, D.: Parallel materialisation of datalog programs in centralised, main-memory RDF systems. In: Proceedings of AAAI. AAAI Press (2014)
13. Nejdl, W.: Recursive strategies for answering recursive queries - The RQA/FQI strategy. In: Proceedings of the 13th International Conference on Very Large Data Bases, pp. 43–50. Morgan Kaufmann Publishers Inc. (1987)
14. Prud'Hommeaux, E., Seaborne, A.: SPARQL query language for RDF. W3C recommendation (2008)
15. Pérez-Urbina, H., Rodriguez-Díaz, E., Grove, M., Konstantinidis, G., Sirin, E.: Evaluation of query rewriting approaches for OWL 2. In: Proceedings of SSWS+HPCSW (2012)
16. Russell, S., Norvig, P.: Artificial Intelligence: A Modern Approach. Prentice Hall, Englewood Cliffs (2009)
17. Salvadores, M., Correndo, G., Harris, S., Gibbins, N., Shadbolt, N.: The design and implementation of minimal RDFS backward reasoning in 4store. In: Antoniou, G., Grobelnik, M., Simperl, E., Parsia, B., Plexousakis, D., De Leenheer, P., Pan, J. (eds.) ESWC 2011, Part II. LNCS, vol. 6644, pp. 139–153. Springer, Heidelberg (2011)
18. Urbani, J., Kotoulas, S., Maassen, J., Van Harmelen, F., Bal, H.: WebPIE: A web-scale parallel inference engine using MapReduce. Web Semant. Sci. Serv. Agents World Wide Web **10**, 59–75 (2012)

19. Urbani, J., Margara, A., Jacobs, C., van Harmelen, F., Bal, H.: DynamiTE: Parallel materialization of dynamic RDF data. In: Alani, H., et al. (eds.) ISWC 2013, Part I. LNCS, vol. 8218, pp. 657–672. Springer, Heidelberg (2013)

20. Urbani, J., Margara, A., Jacobs, C., Voulgaris, S., Bal, H.: AJIRA: A lightweight distributed middleware for MapReduce and stream processing. In: 2014 IEEE 34th International Conference on Distributed Computing Systems (ICDCS), pp. 545–554, June 2014

21. Urbani, J., Piro, R., van Harmelen, F., Bal, H.: Hybrid reasoning on OWL RL. Semant. Web 5(6), 423–447 (2014)

22. Urbani, J., van Harmelen, F., Schlobach, S., Bal, H.: QueryPIE: Backward reasoning for OWL horst over very large knowledge bases. In: Aroyo, L., Welty, C., Alani, H., Taylor, J., Bernstein, A., Kagal, L., Noy, N., Blomqvist, E. (eds.) ISWC 2011, Part I. LNCS, vol. 7031, pp. 730–745. Springer, Heidelberg (2011)

23. Vieille, L.: Recursive Axioms in Deductive Databases: The Query/Subquery Approach. In: Expert Database Conference, pp. 253–267 (1986)

24. Vieille, L.: A database-complete proof procedure based on SLD-resolution. In: Logic Programming, Proceedings of the Fourth International Conference, pp. 74–103 (1987)

25. Vieille, L.: From QSQ towards QoSaQ: global optimization of recursive queries. In: Expert Database Conference, pp. 743–778 (1988)

26. Vieille, L.: Recursive query processing: the power of logic. Theoret. Comput. Sci. 69(1), 1–53 (1989)

27. Weaver, J., Hendler, J.A.: Parallel materialization of the finite RDFS closure for hundreds of millions of triples. In: Bernstein, A., Karger, D.R., Heath, T., Feigenbaum, L., Maynard, D., Motta, E., Thirunarayan, K. (eds.) ISWC 2009. LNCS, vol. 5823, pp. 682–697. Springer, Heidelberg (2009)

28. Zhou, Y., Nenov, Y., Cuenca Grau, B., Horrocks, I.: Complete query answering over horn ontologies using a triple store. In: Alani, H., et al. (eds.) ISWC 2013, Part I. LNCS, vol. 8218, pp. 720–736. Springer, Heidelberg (2013)

29. Zhou, Y., Nenov, Y., Grau, B.C., Horrocks, I.: Pay-as-you-go OWL query answering using a triple store. In: Proceedings of AAAI, pp. 1142–1148. AAAI Press (2014)

Bring User Interest to Related Entity Recommendation

Zhongyuan Wang[1,2]($\boxtimes$), Fang Wang[2,3], Ji-Rong Wen[1], and Zhoujun Li[3]

[1] School of Information, Renmin University of China, Beijing, China
zhy.wang@microsoft.com, jirong.wen@gmail.com
[2] Microsoft Research, Beijing, China
fangwang@buaa.edu.cn
[3] State Key Laboratory of Software Development Environment,
Beihang University, Beijing, China
lizj@buaa.edu.cn

Abstract. Most existing approaches to query recommendation focus on query-term or click based analysis over the user session log or click-through data. For entity query, however, finding the relevant queries from these resources is far from trivial. Entity query is a special kind of short queries that commonly appear in image search, video search or object search. Focusing on related entity recommendation, this paper proposes to collect rich related entities of interest from a large number of entity-oriented web pages. During the collection, we maintain a large-scale and general-purpose related entity network (REN), based upon a special co-occurrence relation between the related entity and target entity. Benefiting from the REN, we can easily incorporate various types of related entity into recommendation. Different ranking methods are employed to recommend related and diverse entities of interest. Extensive experiments are conducted to assess the recommendation performance in term of *Accuracy* and *Serendipity*. Experimental results show that the REN is a good recommendation resource with high quality of related entities. For recommending related entity, the proposed REN-based method achieves good performance compared with a state-of-the-art relatedness measurement and two famous recommendation systems.

Keywords: Query recommendation · Entity ranking · Related entities

1 Introduction

Query recommendation has long been considered a key feature of search engines. Recommending the most relevant queries to users not only enhances the search engine's hit rate, but also helps the user to find the desired information more quickly. However, only considering the relevance may not be enough to guarantee recommendation usefulness and effectiveness [21], especially for entity queries. We call queries that consist of only one entity as entity queries, such as "Tom Cruise," "Facebook," and "America." Entity query is a special kind of short

© Springer International Publishing Switzerland 2015
M. Croitoru et al. (Eds.): GKR 2015, LNAI 9501, pp. 139–153, 2015.
DOI: 10.1007/978-3-319-28702-7_9

queries, which is very common in image/video search or object search. It is non-trivial to capture the user's search intent for an entity query, because it is generally very short and may be ambiguous. In this case, recommending the most *relevant* queries may be redundant. Taking the query "iPhone" for example, recommending "iPhone 4s," "iPhone 5s," and "iPhone 6" may not surprise the user. From the perspective of serendipity, it would be better to recommend queries covering more aspects related to the entity. For example, "iPhone 6," "Samsung Galaxy S5," and "Nokia Lumia 1020." may be good choices for the user.

This paper focuses on related entity recommendation, aiming to provide new entities of interest that are related to various aspects and topics, rather than a set of synonyms or extensions/supplements based on user query. Related entity recommendation is deemed useful. For example, in object-level search engines [18], it is important to return a list of user-interested entities that are highly related to user queries. In addition, it can help users explore wider scope of query intents, so as to inspire users with more searching interests. Figure 1 shows three real recommendation products, from related topics in Bing, related queries in Google and query interpretations in Yahoo!

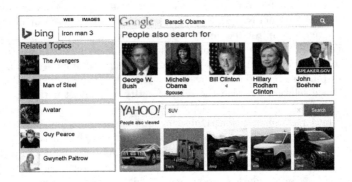

Fig. 1. Examples of entity recommendation products

However, recommending related entities of interest is a challenging task for the following reasons.

- *Query Term Based Analysis may Lead to Redundant Recommendations*: Most existing work [1,12] considers previous queries having common terms with the current query to be the similar queries and naturally recommends these queries. Entity query, however, is meaningful term itself. Term-based recommendation methods may provide surface-similar entities, leading to redundant recommendations.
- *Search Log Based mining methods are limited*: Some recommendation systems [1,6,24] can mine related queries from search log (e.g. session log). For related entity, however, it is non-trivial to collect related entities of interest from previous search log. First, the followed query is more likely to be the

refinement of its former query in the search log, rather than a related query of different topics. Second, the search log itself has many noises. It is hard to identify whether the followed query is a user-interested entity related with its previous query.

– *Related Entities of Interest are Diverse and Abundant*: Users may be interested in various queries because of different reasons. For example, when a user searches for "iPad", he/she may also be interested in other related tablets such as "Surface," or want to learn more about its producer "Apple Inc.," or even associate "iPad" with its founder "Steve Jobs." In addition, entities of interest to users are usually entity-dependent. Therefore, it is hard to define which type of entities is user-interested.

In this paper we present a novel approach for related entity recommendation. It comprises the following two stages.

The first stage aims to gather rich related entities of interest with good coverage in an open domain manner. Instead of using session log or query click log, we collect the related entities by leveraging a large number of entity-oriented web pages, such as Wikipedia articles and product web pages in e-commerce web sites. Our intuition is as follows. Although it is difficult to identify which entities may interest the users, we observe that in an entity description page, users will mention their interested entities that are related to the entity. For example, given "The Amazing Spider-Man 2," we can easily find its related entities that users care about in the movie page[1], such as the *actor* "Andrew Garfield" and its *director* "Marc Webb". Given an entity, we extract its user-interested entities from its description pages.

During this process, a **R**elated **E**ntity **N**etwork (REN) is also maintained based upon a special co-occurrence relationship[2] between entities. As Sangkeun et al. [15] state, exploiting a graph data model for recommendation can easily incorporate various types of information into recommendation, because any type of entities can be modeled as nodes in the graph.

The second stage first employs various methods to study the entity ranking issue. To measure the semantic relatedness between entities, we utilize *semantic analysis* methods based on entity co-neighbors and co-concepts. The conceptual information is introduced by leveraging a large taxonomy knowledgebase. To measure the entity importance, *link analysis* methods are exploited based on the REN, such as *Personalized PageRank* [14].

Diversity is an important feature for recommending related applications, which affects the user experience directly. Most existing approaches [8,10,16] diversify the recommendations based on different aspects or subtopics of the recommended target. Generally, a clustering or other subtopic mining process is needed to generate different aspects. While in our approach, this step is no longer needed, because the proposed ranking method incorporated with concept information can directly gain diverse entities.

---

[1] http://www.movies.com/amazing-spider-man-2/m68476.

[2] The co-occurrence relation limits one of the two entities to be in the title while the other appears in the body text of the same article.

The main contributions of this paper are as follows:

- We design a simple but effective strategy that leverages massive entity-oriented documents to gather rich related entities of interest. A graph data mode - REN is exploited to represent the relation between the entities. Based on the new resource of entity recommendation, we recommend diverse but related entities to users.
- We incorporate various ranking methods to provide relevant entities related to different aspects and topics. Experimental results show that the REN outperforms co-occurrence based approach by 20 % in terms of relatedness between entities, indicating that it is a good recommendation resource. Extensive experiments show that the proposed recommendation method performs well in terms of accuracy and serendipity

2   Related Work

Existing query recommendation techniques differ from one another in terms of the methods they use to find related queries and the techniques they use to rank the candidates for query recommendation.

Baeza-Yates et al. [1] obtain the recommendations from the aggregation of the term-weight vectors of the URLs clicked after the query, using a k-means clustering method. Based on search log, Chien-Kang et al. [12] extracts suggestions from search sessions in a query log that appeared similar to the user's current session. Wen et al. [22] present a density-based clustering method for query recommendation by leveraging the query-click graph [2]. Most of these work considers previous queries having common terms with the current query to be the similar queries. For related entity recommendation, however, the surface-similar queries may lead to redundant recommendations, as the entity queries are generally very short.

Instead, Zhiyong et al. [24] and Qi et al. [11] utilize the sequentiality of the user query to recommend related queries. Boldi et al. [3] generate query suggestions based on random walks in the query-flow graph. They weight the graph edge by leveraging the number of times that the current query was followed by the candidate query. Szpektor et al. [19] propose a template-based method to improve the long-tail query recommendation by leveraging a query-template flow graph. Vahabi et al. [20] propose an orthogonal query recommendation method, which intentionally seeks the related queries that have (almost) no common terms with the user's query, to satisfy the users informational need when small perturbations of the original keyword set are insufficient. Indeed, these methods can mine some related entity of interest from the search log. However, the log itself has many noises. It is hard to identify whether the followed query is a user-interested entity related with its previous query or it is a new search issue that is irrelevant to its former query.

Our method differs from previous approaches in that we find rich entities of interest that are related to various aspects and topics from entity-oriented web pages. In this way, we can recommend new related entities, so as to avoid recommending surface similar queries. Moreover, we utilize the conceptual information from existing knowledgebase to enhance the semantic relatedness between entities, making the recommendations diverse without loss of relevance.

3   Related Entity Collection

In this section, we first describe our strategy for collecting related entity of interest. And then we introduce the construction of the **R**elated **E**ntity **N**etwork (REN).

3.1   Collecting Related Entities

Our construction strategy takes as input a large number of entity-oriented web pages, which are identified with a strict rule: if there is only one entity appearing in the *title* of the web page explicitly. We regard the entity in the *title* of an entity-oriented document as the target entity, and the entities in the body text as the entities related with the target entity. Specifically, it is divided into the following two subtasks:

- *Target Entity Extraction:* We extract the target entity with a wrapper-based strategy following the work by Dalvi, et al. [7]. Note that if the web page is a Wikipedia article, this stage will be skipped, since the title of a Wikipedia article can be regarded as an entity.
- *Related Entity Extraction:* We use Wikipedia Miner [17][3] to identify entities in the text. It is one of the state-of-the-art open domain entity extraction tools, which detects and disambiguates Wikipedia entities when they are mentioned in documents. As we know, linking free text from web pages to existing knowledge bases (e.g., Wikipedia) is far from trivial. We remark that detecting and disambiguating entities in web pages is not the objective of this paper. Improving the entity extraction will probably further improve the quality of extracted user-interested entities. We leave this for future work.

In our strategy, entities are extracted in the form of triples: $< e_i, e_j, n_{ij} >$, where e_i is a target entity, e_j is a related entity and n_{ij} is the co-occurrence frequency that e_j appeared in e_i's pages, for instance, $< iron\ man\ 3, tony, 4 >$. This work uses English Wikipedia dump[4] as the main resource. In addition, several hundreds of domain-specific web sites are also considered. We mainly focus on web sites that are full of entities, such as product web sites like Amazon.com (book, electronic products, etc.), Music web sites, and Movie web sites. In total, we obtain 3.9 million entities and 77 million relations between them.

---

[3] http://wikipedia-miner.sourceforge.net.
[4] This work uses enwiki dump progress on 20130503.

3.2 Constructing the REN

The REN Graph can be represented as a directed graph $G_U = (V, E, \varphi)$ where:

- $V = \{e_1, e_2, \cdots, e_n\}$ is the set of entities in the graph;
- $E \subseteq V \times V$ is the set of *directed* edges.
- $\varphi\colon E \to R_+$ is a weighting function which assigns a weight w to each edge, where R_+ is the set of positive real numbers.

In our setting, a node in the graph represents an entity. An edge $< e_i, e_j >$ represents that e_i is the target entity and e_j is its related entity. The weight on each edge indicates the relatedness strength between the two connected entities, which plays an important role in ranking the related entities in the REN.

Figure 2 shows a small fragment of the REN. In this figure, the weight on each edge is measured with *tf-idf*. In our case, tf is the frequency of e_j appearing in e_i' pages and *idf* is the inverse document frequency of e_j where we regard each entity-oriented page as a document. As can be seen from the fragment, our collection strategy can gather many user-interested entities of various types, such as *Person* ("Steve Jobs"), *Company* ("Apple Inc."), and *Device* ("iPhone 4s"). For the entity "iPad", its user-cared entities include "Apple Inc.", "Touchscreen", "Wireless", "Tablet", "Apple store", etc. We can also see that the collected entities are highly related with their target entities.

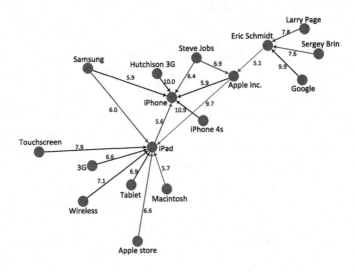

Fig. 2. A fragment of the constructed REN Graph

4   Related Entity Ranking

For ranking the related entities, *co-occurrence frequency* and *tf-idf* are two naive methods. However, it may not work well in measuring the relevance because they make no use of semantic similarities between entities. In this section, we discuss various possible ranking functions to study the relatedness ranking issue.

To capture more semantic information of entities, we assign a set of concepts (e.g., *Device*, *Company* and *Person*) from Probase [23] to each entity in the REN. To the best of our knowledge, Probase[5] is the largest general-purpose taxonomy, containing 2.7 M concepts (e.g. *USA President*) and 40 M entities (e.g. "Barack Obama"). It also contains 4.5 M *Is-a* relations, such as *"robin" is-a bird*. Moreover, Probase gives $p(c|e)$, the *typicality* of concept c among all concepts that contain instance e, to measure the probability of an entity belonging to a concept [23]. In our case, for each entity, we rank its concepts by typicality and select the top 20 concepts as their concept level information, which in most cases can cover all the concepts of an entity.

4.1   Ranking by Semantic Relatedness

We study the semantic relatedness between the target entity and its related entity based on two factors: concepts and neighboring entities.

As we know that entities belonging to similar concepts are semantically similar to each other. Thus, we represent the entities by vectors of Probase concepts, and then measure the relatedness by comparing the corresponding vectors using conventional metrics. Two famous similarity measurements are used here:

– *Concept Overlap:* this is Jaccard similarity on co-concepts of e_i and e_j;

$$C_{overlap}(e_i, e_j) = \frac{|C(e_i) \cap C(e_j)|}{|C(e_i) \cup C(e_j)|} \qquad (1)$$

where $C(e_i)$ is the concept set of e_i.
– *Concept Similarity:* this is cosine similarity based on entity concepts.

$$Sim_c(e_i, e_j) = \frac{\sum C(e_i) \times C(e_j)}{\sqrt[2]{\sum C^2(e_i)} \times \sqrt[2]{\sum C^2(e_j)}} \qquad (2)$$

The above two methods measure the concept level similarity between two entities. E.g., $Sim_c(China, India) = 0.99$ indicates that mostly "China" and "India" belong to the same concepts (e.g., *Country*, *Developing Country* and *Asia Country*). From the aspect of neighboring entities, we believe that entities with similar neighbors are semantically similar. Same as above, two similar measurements are used.

– *Neighborhood Overlap:* it could be seen as Jaccard similarity on co-neighbor entities of e_i and e_j;

$$E_{overlap}(e_i, e_j) = \frac{|Nei(e_i) \cap Nei(e_j)|}{|Nei(e_i) \cup Nei(e_j)| - 2} \qquad (3)$$

where $Nei(e_i)$ is the set of e_i' neighbors.

---

[5] Probase data is publicly available at http://probase.msra.cn/dataset.aspx.

- *Similarity Based on Entity Neighbors:* this is cosine similarity based on neighbored entities.

$$Sim_e(e_i, e_j) = \frac{\sum Nei(e_i) \times Nei(e_j)}{\sqrt[2]{\sum Nei^2(e_i)} \times \sqrt[2]{\sum Nei^2(e_j)}}. \tag{4}$$

4.2   Ranking by Entity Importance

Apart from the semantic relatedness, we believe that the popularity or importance of an entity is also an important feature for entity recommendation. For example, given the target entity "iPhone," it would be better to recommend "Steven Jobs" than "Hutchison 3G" because the former is a more famous entity to users. This famous or popular feature can be captured by link analysis base on the REN.

In this paper, we adapt *Personalized PageRank* [14] for ranking the entity importance with respect to the target entities. Given an entity query e_i, we define its personalized related entities as those that have co-entity or co-concept with the target entity. Formally, we define its $|V| \times 1$ personalized vector $\widetilde{\mathbf{v}}$ as follows:

$$\mathbf{v}_j = \begin{cases} 1 & : & if \ E_{overlap}(e_i, e_j) + C_{overlap}(e_i, e_j) > 0 \\ 0 & : & otherwise \end{cases} \tag{5}$$

We use the normalized vector $\mathbf{v}$ of $\widetilde{\mathbf{v}}$ as the personalized Pagerank vector (PPV) for ranking entities on the REN. Formally, the Personalized PageRank score can be calculated as:

$$\mathbf{r} = (1 - \alpha)\mathbf{Mr} + \alpha\mathbf{v} \tag{6}$$

where $\mathbf{M}$ is a $|V| \times |V|$ conductance matrix. The element (e_i, e_j) is conductance $Pr(e_i|e_j)$, i.e., the probability that the random surfer walks from e_j to e_i. This paper uses the normalized co-occurrence frequency between e_i and e_j to estimate $Pr(e_i|e_j)$. $0 < \alpha < 1$ is the probability of walking (as against jumping) in each step.

We also use *PageRank* [5] to measure the global importance of an entity in the REN.

5   Experiment

In this section, we evaluate the proposed approach from the following aspects: REN quality evaluation, ranking method evaluation, relatedness comparison and recommendation comparison. The evaluation dataset is introduced first.

5.1   Dataset

We test the performance of our method for mining and ranking related entities by using a set of test entities. Ten target entities are selected from ten general

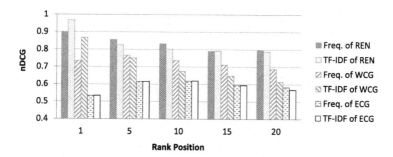

Fig. 3. The REN quality evaluation

categories: "China" for *Geography*, "Tang Dynasty" for *History*, "Psychology" for *Society*, "Microsoft" for *Technology*, "Dollar" for *Economy*, "Chocolate" for *Daily Life*, "Gone with the wind" for *Literature & Art*, "Baseball" for *Sports*, "Warren Buffett" for *People* and "SUV" for *Cars*. For each entity, we select its neighboring entities in the REN as the candidate entities. Thus, we get 1728 entity pairs, more than 100 related ones per target entity, which constitute our experimental data.

We organize the target entities and their related entities as entity pairs. We recruit three human judges to label the relevance of the candidate entities regarding the target entity. The label criteria is: Label 0 for *irrelevant* entity pair, 1 for *not sure*, 2 for *related* and 3 for highly related. Each entity pair gets three labels and the majority of the labels is taken as the final decision for an entity pair. If all three annotators do not agree with each other, the arithmetic average of the assigned label values will be used.

5.2 The REN Quality Evaluation

To assess the quality of related entities in the REN, we compare it with the following two entity co-occurrence graphs:

– *Wikipedia Co-occur Graph (WCG)*: We use the same English Wikipedia dump[6] as the REN to build a Wiki entity co-occur graph, based on document-wise co-occurrence. That is, given a target entity, its related entities are those that co-occurred with it in wiki articles.
– *Entity Co-occur Graph (ECG)*: We also construct an entity co-occurrence graph from the Web as a baseline method. The data was collected from sentences in 1% of search log snapshots from a popular search engine, which is kind of sentence-wise co-occurrence. We identify the entity in each sentence by using Freebase [4] as a lexicon of entities.

To make a fair comparison, we use the same method to select and label the two datasets. Since we only focus on the data quality in current comparison, two simple weighting schemas: *co-occurrence frequency* and *tf-idf* are used

---

[6] This work uses enwiki dump progress on 20130503.

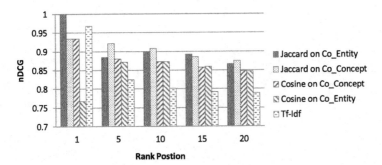

(a) Ranking results based on semantic relatedness

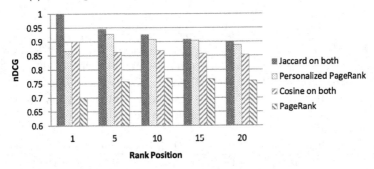

(b) Ranking results based on entity importance

Fig. 4. Related entity ranking with various measures

for ranking the related entities in the three datasets. We then use the $nDCG$ (Normalized Discounted Cumulative Gain) to assess the relevance of the top 20 results retrieved for the ten target entities.

Experimental results are respectively plotted in Fig. 3. Both the ranking methods show that the relatedness precision of the REN is significantly higher than those of the two co-occurrence data. On average, the performance of the REN in terms of entity relatedness is 9.82 % higher than Wikipedia co-occur data and much better than Web entity co-occur data (21.77 %). This shows the high quality of related entities in the REN, making it a good resource for entity recommendation.

5.3   Performance of Different Ranking Methods

We also use $nDCG$ as the evaluation metric. The ten selected entities are regarded as entity queries. In this case, for each target entity, we randomly select 20 related entities to constitute a ranking group. Finally, we obtain 81 ranking groups that used as our test dataset.

For each entity pair, we compute four semantic scores marked as: Sim_{J_c} by Eq. (1), Sim_{J_e} by Eq. (3), Sim_{C_c} by Eq. (2) and Sim_{C_e} by Eq. (4). Figure 4(a) shows the ranking results.

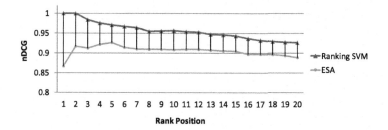

Fig. 5. Overall related entity ranking performance

We also use the proposed *Personalized PageRank* (Eq. 6) and the traditional *PageRank* to compute the entity importance, where the α is set to 0.85. For better comparison with the above semantic ranking, we simply add Sim_{J_c} and Sim_{J_e} as a combined *Jaccard* score based on co-concept and co-neighbor, marked as *Jaccard on Both*. The combined *cosine* score is computed in the same way, marked as *Cosine on Both*. Figure 4(b) shows the ranking results.

From this figure, we draw the following observations:

(1) As can be seen from Fig. 4(a), compared with *tf-idf*, the ranking precision is greatly improved by using semantic similarity measures, with the accuracy enhanced by at least 5 %.
(2) Adding concept information can improve the ranking performance. Better results are achieved by combining the semantic information of co-concept and co-neighbor. The best accuracy is achieved by the combined *Jaccard* score, as high as 90 % even at rank 20.
(3) For ranking the related entities in the REN, the measurements based on link analysis are inferior to semantic relatedness based measurements. *Personalized PageRank* performs much better than the traditional *PageRank*. We think this is because it considers more semantic information through the personalized vector (Eq. 5).

5.4   Relatedness Ranking Comparison

To assess the overall performance of the related entity ranking, we employ Ranking SVM as a combined ranking method. In this comparison, the six ranking methods are used as features to train Ranking SVM. We randomly split the labeled data into two data sets, one for training the Ranking SVM and the other for comparison with ESA. This work uses SVM^{rank} [13] for efficiently training Ranking SVM. Ranking SVM assigns a score to each candidate entity. The higher the score, the better the candidate is as a user-interested entity. Similarly, each candidate entity also gets a relatedness score from ESA. Figure 5 shows the experimental results.

The performance of our related entity ranking is comparable with that of ESA, with the average $nDCG$ higher by 5.1 %. This demonstrates that the proposed ranking methods are effective in ranking the related entities in terms of relatedness. Through a deeper analysis of their ranking results (Table 1), we notice that the results ranked by our combined method are not only highly related but also diverse and consistent with human thinking, due to the semantic ranking methods.

5.5 Entity Recommendation Comparison

In this subsection, we assess the recommendation performance through a comparison with two famous recommendation systems.

Metric. Serendipity takes into account the novelty of recommendations and how much recommendations may positively surprise users (or inspirit user more searching interests). We adapt metric $SRDP$ in Eq. 7 to evaluate the performance of the combined ranking method in terms of serendipity. Ge et al. [9] designed this metric to capture the unexpectedness and relevance of the recommendations.

$$SRDP(RS) = \frac{\sum_{i \in UNEXP} rel(i)}{|UNEXP|} \tag{7}$$

where RS denotes the set of recommendations generated by a recommender system, $UNEXP$ represents the set of unexpected recommendations in RS that are not in a baseline recommendation list. The function $rel(i)$ is used to evaluate the relevance of a recommendation.

Baselines. We use two baseline recommender systems: Bing image search and Yahoo! image search. For the ten target entities, we first query the target entity

Table 1. Examples of the ranking results (The top 5)

Query	Ranking SVM	ESA
Microsoft	Google	Microsoft publisher
	Novell	Microsoft office
	Sony	Microsoft project
	Microsoft office	Microsoft infopath
	Nokia	Microsoft office
Psychology	Sigmund freud	Free association
	Stanley milgram	Media psychology
	Philosophy	Social psychology
	Clinical psychology	Clinical psychology
	Social science	Pleasure principle
SUV	Jeep wagoneer	Off-road vehicle
	Sedan	Recreational vehicle
	Gl-class	General motors
	Cadillac escalade	American motors
	Lexus lx	Recreational vehicle

through the search interface and then collect the top 5 recommendations as the baseline data sets. For relevance, we use ESA as well as the judgements used in the dataset labeling. In this case, we use the combined method (please refer to the previous subsection) to generate the recommendations from the REN for each target entity. To assess the performance, we compare our recommendations with those that are collected from Bing and Yahoo! separately.

Results. Experimental results are given in Table 2. We abbreviate *recommendations* to *recs* due to the space limitation. In comparison with Bing, we use the *recs* generated by Yahoo! as the baseline. As shown in the first row, the serendipity of the *recs* from the REN is 0.83. This score denotes that the unexpected *recs* of the REN ranking, which are not in Yahoo!'s recommendation list, are highly related to the target entities under the evaluation of human judgements. We can see that our REN ranking outperforms Bing and Yahoo! in both of the two evaluation methods. This indicates that the REN ranking can positively surprise users with more related recommendations. Note the serendipity computed by ESA is much lower than that from human judgements, because the relevance score from ESA is much lower than the human labeling score. On the whole, it is consistent with the human judge.

Table 2. Serendipity evaluation results, by using two relevance functions (Human Judgement and ESA)

Baseline	Recs from	Human judgement	ESA
Yahoo!	REN ranking	**0.83**	**0.27**
	Bing	0.76	0.26
Bing	REN ranking	**0.84**	**0.28**
	Yahoo!	0.73	0.12

We have transferred the REN ranking features to Bing on related entity recommendation. Now it is successfully used in Bing image search. The feedback is pretty good. Both the Distinct Search Queries (DSQ) rate and the Click Through Rate (CTR) have achieved significant gains.

6   Conclusion

This paper proposes a novel approach for related entity recommendation. Inspired by the fact that many user-cared entities are mentioned in entity-oriented pages, we mine rich entities related to various aspects and topics from entity-oriented web pages, instead of search engine log. Massive related entities of interest are collected automatically from a large number of entity-oriented pages. During the process, a large-scale, general-purpose, related entity network (REN) is maintained based upon the co-occurrence relationship. We employ various related entity ranking methods based on the REN, including co-concept and co-entity based semantic similarity methods and graph-based link analysis methods. Extensive experiment is conducted to show the high quality of related entities in the REN. We compare our method to existing famous recommendation systems. The results show that our method achieves better top-5 recommendation performances in terms of serendipity. Also, we show that exploiting conceptual information from existing knowledge bases can improve recommendation quality.

Acknowledgments. This work was partially supported by the National Key Basic Research Program (973 Program) of China under grant No. 2014CB340403 and the Fundamental Research Funds for the Central Universities & the Research Funds of Renmin University of China. This work was also supported in part by NSFC (Nos. 61170189, 61370126, 61202239), National High Technology Research and Development Program of China under grant No.2015AA016004, the Fund of the State Key Laboratory of Software Development Environment (No. SKLSDE-2015ZX-16), and Microsoft Research Asia Fund (No. FY14-RES-OPP-105).

References

1. Baeza-Yates, R., Hurtado, C.A., Mendoza, M.: Query recommendation using query logs in search engines. In: Lindner, W., Fischer, F., Türker, C., Tzitzikas, Y., Vakali, A.I. (eds.) EDBT 2004. LNCS, vol. 3268, pp. 588–596. Springer, Heidelberg (2004)
2. Beeferman, D., Berger, A.: Agglomerative clustering of a search engine query log. In: SIGKDD, pp. 407–416. ACM (2000)
3. Boldi, P., Bonchi, F., Castillo, C., Donato, D., Gionis, A., Vigna, S.: The query-flow graph: model and applications. In: CIKM, pp. 609–618. ACM (2008)
4. Bollacker, K., Evans, C., Paritosh, P., Sturge, T., Taylor, J.: Freebase: a collaboratively created graph database for structuring human knowledge. In: SIGMOD, pp. 1247–1250. ACM (2008)
5. Brin, S., Page, L.: The anatomy of a large-scale hypertextual web search engine. Comput. Netw. ISDN Syst. **30**(1), 107–117 (1998)
6. Cao, H., Jiang, D., Pei, J., He, Q., Liao, Z., Chen, E., Li, H.: Context-aware query suggestion by mining click-through and session data. In: SIGKDD, pp. 875–883. ACM (2008)
7. Dalvi, N., Kumar, R., Soliman, M.: Automatic wrappers for large scale web extraction. VLDB **4**(4), 219–230 (2011)
8. Dang, V., Croft, W.B.: Diversity by proportionality: an election-based approach to search result diversification. In: SIGIR, pp. 65–74. ACM (2012)
9. Ge, M., Delgado-Battenfeld, C., Jannach, D.: Beyond accuracy: evaluating recommender systems by coverage and serendipity. In: RecSys, pp. 257–260. ACM (2010)
10. He, J., Hollink, V., de Vries, A.: Combining implicit and explicit topic representations for result diversification. In: SIGIR, pp. 851–860. ACM (2012)
11. He, Q., Jiang, D., Liao, Z., Hoi, S.C., Chang, K., Lim, E.-P., Li, H.: Web query recommendation via sequential query prediction. In: ICDE, pp. 1443–1454. IEEE (2009)

12. Huang, C.-K., Chien, L.-F., Oyang, Y.-J.: Relevant term suggestion in interactive web search based on contextual information in query session logs. JASIST **54**(7), 638–649 (2003)
13. Joachims, T.: Training linear svms in linear time. In: KDD, pp. 217–226. ACM (2006)
14. Kleinberg, J.M.: Authoritative sources in a hyperlinked environment. JACM **46**(5), 604–632 (1999)
15. Lee, S., Song, S.-I., Kahng, M., Lee, D., Lee, S.-G.: Random walk based entity ranking on graph for multidimensional recommendation. In: RecSys, pp. 93–100. ACM (2011)
16. Li, R. Kao, B., Bi, B., Cheng, R., Lo, E.: Dqr: a probabilistic approach to diversified query recommendation. In: CIKM, pp. 16–25. ACM (2012)
17. Milne, D., Witten, I.H.: An open-source toolkit for mining wikipedia. Artif. Intell. **194**, 222–239 (2013)
18. Nie, Z., Wen, J.-R., Ma, W.-Y.: Object-level vertical search. In: CIDR, pp. 235–246 (2007)
19. Szpektor, I., Gionis, A., Maarek, Y.: Improving recommendation for long-tail queries via templates. In: WWW, pp. 47–56. ACM (2011)
20. Vahabi, H., Ackerman, M., Loker, D., Baeza-Yates, R., Lopez-Ortiz, A.: Orthogonal query recommendation. In: RecSys, pp. 33–40. ACM (2013)
21. Vargas, S., Castells, P.: Rank and relevance in novelty and diversity metrics for recommender systems. In: RecSys, pp. 109–116. ACM (2011)
22. Wen, J.-R., Nie, J.-Y., Zhang, H.-J.: Clustering user queries of a search engine. In: WWW, pp. 162–168. ACM (2001)
23. Wu, W., Li, H., Haixun, W., Zhu, K.Q.: Probase: a probabilistic taxonomy for text understanding. In: SIGMOD, pp. 481–492. ACM (2012)
24. Zhang, Z., Nasraoui, O.: Mining search engine query logs for query recommendation. In: WWW, pp. 1039–1040. ACM (2006)

Author Index

Printed in the United States
By Bookmasters